ANIMAL TRACKS

of

ARIZONA &
NEW MEXICO

IAN SHELDON

LONE PINE

© 1998 by Lone Pine Publishing
First printed in 1998   10 9 8 7 6 5 4 3 2

Printed in Canada

THE PUBLISHER: LONE PINE PUBLISHING

10145 – 81 Avenue
Edmonton, AB T6E 1W9
Canada

1808 – B Street NW, Suite 140
Auburn, WA 98007
USA

Lone Pine Publishing website: http://www.lonepinepublishing.com

Canadian Cataloguing in Publication Data

Sheldon, Ian

Animal tracks of Arizona & New Mexico
Includes bibliographical references and index.
ISBN 13 978-1-55105-145-1
ISBN 10 1-55105-145-1

1. Animal tracks–Arizona–Identification. 2. Animal tracks–New
Mexico–Identification. I. Title.
QL768.S519 1998   591.47'9   C98-910840-6

Senior Editor: Nancy Foulds
Editor: Volker Bodegom
Production Manager: David Dodge
Design, layout and production: Volker Bodegom, Gregory Brown
Map: Volker Bodegom
Technical review: Donald L. Pattie
Animal illustrations: Gary Ross, Horst Krause, Kindrie Grove
Track illustrations: Ian Sheldon
Cover illustrations: White-tailed Jackrabbit and Pronghorn Antelope by Gary Ross
Scanning: Elite Lithographers Ltd., Edmonton, Alberta, Canada

The publisher gratefully acknowledges the support of Alberta Community
Development and the Department of Canadian Heritage.

PC: P1

CONTENTS

INTRODUCTION

If you have ever spent time with an experienced track-er, or perhaps a veteran hunter, then you know just how much there is to learn about the subject of tracking and just how exciting the challenge of tracking animals can be. Maybe you think that tracking is no fun, because all you get to see is the animal's prints. What about the animal itself–isn't that much more exciting? Well, for most of us who don't spend a great deal of time in the beautiful wil-derness of the US southwest, the chances of seeing the graceful Pronghorn at full speed or spotting the elusive Mountain Lion are slim. The closest we may ever get to some animals will be through their tracks, and they can inspire a very intimate experience. Remember, you are following in the footsteps of the unseen–animals that are in pursuit of prey, or perhaps being pursued as prey.

This book offers an introduction to the complex world of tracking animals. Sometimes tracking is easy. At other times it is an incredible challenge that leaves you wonder-ing just what animal left those unusual tracks. Take this book into the field with you, and it can provide some help with the first steps to identification. Prints and tracks are this book's focus; you will learn to recognize subtle differ-ences for both. There are, of course, many additional signs to consider, such as scat and food caches, all of which help you to understand the animal that you are tracking.

Remember, it takes many years to become an expert tracker. Tracking is one of those skills that grows with you as you acquire new knowledge in new situations. Most importantly, you will have an intimate experience with nature. You will learn the secrets of the seldom seen. The more you discover, the more you will want to know, and by developing a good understanding of tracking, you will gain an excellent appreciation of the intricacies and delights of our marvellous natural world.

How to Use this Book

Most importantly, take this book into the field with you! Relying on your memory is not an adequate way to identify tracks. Track identification has to be done in the field, or with detailed sketches and notes that you can take home. Much of the process of identification is circumstantial, so you will have much more success when standing beside the track.

This book is laid out so as to be easy to use. There is a quick reference appendix to the tracks of all the animals illustrated in this book beginning on p. 139. This appendix is a fast way to familiarize yourself with certain tracks and the content of this book, and it guides you to the more informative descriptions of each animal and track.

Each animal's description is illustrated with the appropriate footprints and the styles of track that it usually leaves. While these illustrations are not exhaustive, they do show the tracks or groups of prints that you will most likely see. Where there are differences in orientation, left prints are illustrated. You will find a list of dimensions for the tracks, giving the general range, but there will always be extremes, just as there are with people who have unusually small or large feet. Under the category 'Size' (of animal), the 'greater-than' sign (>) is used when the size difference between the sexes is pronounced.

If you think that you may have identified a track, check the 'Similar Species' section for that animal. This section is designed to help you confirm your conclusions by pointing out other animals that leave similar tracks and showing you ways to distinguish among them.

As you read this book, you will notice an abundance of words such as 'often,' 'mostly' and 'usually.' Unfortunately, tracking will never be an exact science; we cannot expect animals to conform to our expectations, so be prepared for the unpredictable.

Tips on Tracking

As you flip through this guide, you will notice clear, well-formed prints. Do not be deceived! It is a rare track that will ever show so clearly. For a good, clear print, the perfect conditions are slightly wet, shallow snow that isn't melting, or slightly soft mud that isn't actually wet. Needless to say, these conditions can be rare—most often you will be dealing with incomplete or faint prints, where you cannot even really be sure of the number of toes.

Should you find yourself looking at a clear print, then the job of identification is much easier. There are a number of key features to look for: Measure the length and width of the print, count the number of toes, check for claw marks and note how far away they are from the body of the print, and look for a heel. Keep in mind other, more subtle features, such as the spacing between the toes and whether they are parallel or not, and whether fur on the sole of the foot has made the print less clear.

When you are faced with the challenge of identifying an unclear print—or even if you think that you have made a successful identification from one print alone—look beyond the single footprint and search out others. Do not rely on the dimensions of one print alone, but collect measurements from several prints to get an average impression. Even the prints within one track can show a lot of variation.

Try to determine which is the fore print and which is the hind, and remember that many animals are built very differently from humans, having larger forefeet than hind feet. Sometimes the prints will overlap, or they can be directly on top of one another in a direct register. For some animals, the fore and hind prints are pretty much the same.

Check out the pattern that the prints make together in the track, and follow the trail for as many paces as is necessary for you to become familiar with the pattern. Patterns are very important, and can be the distinguishing feature between different animals with otherwise similar tracks.

Follow the trail for some distance, because it may give you some vital clues. For example, the trail may lead you to a tree, indicating that the animal is a climber, or it may lead down into a burrow. This part of tracking can be the most rewarding, since you are following the life of the animal as it hunts, runs, walks, jumps, feeds or tries to escape a predator.

Take into consideration the habitat. Sometimes very similar species can be distinguished only by their locations—one might be found on the riverbank, while another might be encountered only in the dense forest.

Think about your geographical location, too. Some animals have a limited range—perhaps they are found only in remote parts of the Colorado Plateau or are confined to the deserts of Arizona and Nevada. This consideration can rule out some species and help you with your identification.

Remember that every animal will at some point leave a print or track that looks just like the print or track of a completely different animal!

Lastly, keep in mind that if you track quietly, you might catch up with the maker of the prints.

9

Terms and Measurements

Some of the terms used in tracking can be rather confusing, and they often depend on personal interpretation. For example, what comes to your mind if you see the word 'hopping'? Perhaps you see a person hopping about on one leg, or perhaps you see a rabbit hopping through the countryside. Clearly, one person's perception of motion can be very different from another's. Some useful terms are explained below, to clarify what is meant in this book, and where appropriate, how the measurements given fit in with each term.

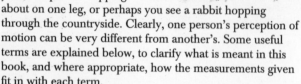

The following terms are sometimes used loosely and interchangeably–for example, a rabbit might be described as 'a hopper' and a squirrel as 'a bounder,' yet both leave the same pattern of prints in the same sequence.

Bounding: Can be used interchangeably with 'hopping' or 'jumping'; often used for patterns that cover short distances; hind prints usually registering ahead of fore prints.

Galloping: Used for the motion made by animals with four even legs, such as dogs, moving at speed; hind prints registering ahead of fore prints.

Hopping: Similar to bounding; usually indicated by tight clusters of prints; fore prints set between and behind the hind prints.

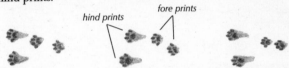

Running: Like galloping, but often applied generally to animals moving at speed.

Stotting (applies to Mule Deer only): Describes the action of taking off from the ground and landing on all four feet at once, in pogo-stick fashion.

Trotting: Faster than walking, slower than running.

Other Tracking Terms

Alternating track: A left-right sequence, as made by humans walking; often a double register for four-legged animals, which are described as 'diagonal walkers.'

double register   direct register

Dewclaws: Two small, toe-like structures set above and behind the main foot of most hoofed animals.

dewclaw marks

dragline

Direct register: The hind print falls directly on the fore print.

Double register: The hind print overlaps the fore print only slightly or falls beside it, so that both can be seen at least in part.

Draglines: The lines left in snow or mud by a foot or the tail dragging over the surface.

Gallop group: A cluster of four prints made at a gallop, usually with hind prints registering ahead of fore prints.

Height: Taken at the animal's shoulder.

Length: The animal's body length from head to rump, not including the tail, unless otherwise indicated.

Lope: A collection of four prints made at a fast pace, usually falling roughly in a line.

Print: Fore and hind prints are treated individually; print dimensions are 'length' (including claws–maximum values may represent occasional heel register for some animals) and 'width'; together the prints make up a track.

Register: To leave a mark–said about a foot, claw or other part of an animal's body.

Retractable: Describes claws that can be pulled in to keep them sharp, as with the cat family; these claws do not register in the prints.

Sitzmark: The mark left on the ground by animal falling or jumping from a tree.

Straddle: The total width of the track, all prints considered.

Stride: For consistency among different animals, the stride is taken as the distance from the centre of one print (or print group) to the centre of the next one. Other books may use the term 'pace.'

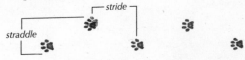

stride

straddle

Track: A pattern left by a series of prints.

Trail: Often used to describe a track at length; think of it as the path of the animal.

MAMMALS

Elk

Fore and Hind Prints
Length: 3.2–5.0 in (8.1–13 cm)
Width: 2.5–4.5 in (6.4–11 cm)

Straddle
7.0–12 in (18–30 cm)

Stride
Walking: 16–34 in (41–86 cm)
Galloping: 3.3–7.8 ft (1.0–2.4 m)
Group length: to 6.3 ft (1.9 m)

Size (stag>hind)
Height: 4.0–5.0 ft (1.2–1.5 m)
Length: 6.5–10 ft (2.0–3.0 m)

Weight
500–1000 lb (230–450 kg)

gallop print   *walking*

ELK (Wapiti)
Cervus elaphus

The Elk, largest of the region's deer, had its once wide range much reduced. Thanks to successful reintroductions, sighting are most likely in northern Arizona and northern New Mexico, and into Utah and Colorado. Female Elk and young are often seen in social herds as they feed in forest openings and lush meadows. Stags, who prefer to go solo, are easily recognized by their magnificent racks of antlers and distinctive bugling in late August. A good place to look for Elk tracks is in the mud beside summer ponds, where Elk like to drink and sometimes splash around.

Elk leave a neat, alternating track with large, rounded prints, often in well-worn winter paths. The hind print will sometimes double register slightly ahead of the fore print. In deeper snow, or if an Elk gallops (with its toes spread wide) the dewclaws may register.

Similar Species: Smaller deer (pp. 18–21) leave similar tracks, but with generally shorter, narrower prints.

Mule Deer

Fore and Hind Prints
Length: 2.0–3.3 in (5.1–8.4 cm)
Width: 1.6–2.5 in (4.1–6.4 cm)
Straddle
5.0–10 in (13–25 cm)
Stride
Walking: 10–24 in (25–61 cm)
Stotting: 9.0–19 ft (2.7–5.8 m)
Size (buck>doe)
Height: 3.0–3.5 ft (91–110 cm)
Length: 4.0–6.5 ft (1.2–2.0 m)
Weight
100–450 lb (45–200 kg)

walking

stot group

MULE DEER
(Black-tailed Deer)
Odocoileus hemionus

The widespread Mule Deer is frequently seen anywhere from mountain meadows and open woodlands to the arid plains of higher elevations, but not on the plains of eastern New Mexico. In winter it moves down from higher terrain to warmer south-facing slopes and sagebrush flats, where it can feed without having to contend with deep snow.

The Mule Deer has a neat, alternating walking track, with the hind print registering on the fore. In winter, this deer prefers to stay in small groups and frequently uses the same well-worn trail. Mule Deer prints are heart-shaped and sharply pointed. In deeper snow or when a deer is moving quickly in mud, its prints show dewclaws, which are closer to the toes on the fore print. At speed, these deer jump with all their feet leaving and striking the ground at once: 'stotting.' Stotting tracks show how the toes spread to distribute the weight and give better footing.

Similar Species: White-tailed Deer (p. 20) prefer forest. Antelope (p. 22) prints have a wider base. Elk (p. 16) have longer, wider prints. Feral Pigs (p. 26) usually register large dewclaws. Peccary (p. 28) prints are much smaller.

White-tailed Deer

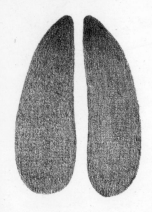

Fore and Hind Prints
Length: 2.0–3.5 in (5.1–8.9 cm)
Width: 1.6–2.5 in (4.1–6.4 cm)
Straddle
5.0–10 in (13–25 cm)
Stride
Walking: 10–20 in (25–51 cm)
Jumping: 6.0–15 ft (1.8–4.6 m)
Size (buck>doe)
Height: 3.0–3.5 ft (91–110 cm)
Length: to 6.3 ft (1.9 m)
Weight
120–350 lb (54–160 kg)

walking   *gallop group*

WHITE-TAILED DEER
Odocoileus virginianus

The keen eyesight of this deer guarantees that it knows about you before you know about it. Frequently, all that we see is its conspicuous white tail as it gallops away, which earns this deer the nickname 'flagtail.' White-tailed Deer may be found in small groups at the edges of forests and in brushlands, in scattered populations primarily in New Mexico, southern Arizona and eastern Colorado.

White-tailed Deer prints are heart-shaped and pointed. The alternating walking track shows the hind prints direct or double registering on the fore prints. In snow or when a deer gallops on soft surfaces, the dewclaws register. This flighty deer gallops in the usual style, hind prints in front of fore prints, with toes spread wide for better footing.

Similar Species: Mule Deer (p. 18) make similar prints but prefer more open terrain and stot (not gallop) at speed. Elk (p. 16) prints are longer and wider. Antelope (p. 22) prefer open spaces. Feral Pig (p. 26) prints usually show prominent dewclaws. Peccary (p. 28) prints are much smaller.

Pronghorn Antelope

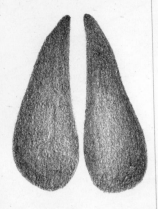

Fore and Hind Prints
Length: 3.3 in (8.4 cm)
Width: 2.5 in (6.4 cm)

Straddle
3.5–9.0 in (8.9–23 cm)

Stride
Walking:
  8.0–19 in (20–48 cm)
Galloping:
  14 ft (4.3 m) or more

Size (buck>doe)
Height: 3.0 ft (91 cm)
Length: 3.8–4.9 ft (1.2–1.5 m)

Weight
75–130 lb (34–59 kg)

walking　　　*gallop group*

PRONGHORN ANTELOPE
Antilocapra americana

This gracious antelope frequents the wide-open grasslands and plains throughout this region. Pronghorns gather in groups of up to a dozen animals in summer and as many as one hundred in winter; they prefer to feed in areas where the snow has blown away. The Pronghorn, one of the fastest animals in North America, runs for fun, easily attaining constant speeds of 40 mph (65 km/h) and short bursts of up to 60 mph (95 km/h).

A Pronghorn print has a pointed tip and a broad base. This animal does not have dewclaws. The hind prints usually register directly on top of the fore prints, making a tidy, alternating track. Pronghorns tend to drag their feet in snow. During their frequent gallops, their toe tips spread wide. The faster the antelope moves, the greater the distance between gallop groups.

Similar Species: Deer (pp. 16–21) prints may show dewclaws and have narrower toe bases; Mule Deer often make stotting tracks. Feral Pig (p. 26) prints often show large dewclaws. Also compare with Collared Peccary (p. 28) prints.

23

Bighorn Sheep

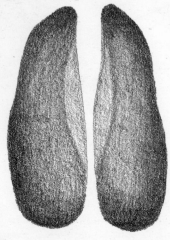

Fore and Hind Prints
Length: 2.5–3.5 in (6.4–8.9 cm)
Width: 1.8–2.5 in (4.6–6.4 cm)

Straddle
6.0–12 in (15–30 cm)

Stride
Walking: 14–24 in (36–61 cm)

Size (ram>ewe)
Height: 2.5–3.5 ft (76–110 cm)
Length: 4.0–6.5 ft (1.2–2.0 m)

Weight
75–270 lb (34–120 kg)

walking

BIGHORN SHEEP
(Mountain Sheep, Desert Bighorn)
Ovis canadensis

In late fall, the loud crack of two majestic rams head-butting one another can be heard for a great distance. To watch the rut is an awe-inspiring experience. Found in the barren lands of desert country, this sheep prefers high, arid terrain and mountain slopes, but moves into valleys during winter. It does not occur east of New Mexico's Sacramento Mountains and it tends to avoid forested areas.

The print is quite squarish in shape and pointed towards the front. The outer edge of the hoof is hard, while the inner part is soft, giving the sheep a good grip on tricky terrain. Find one track and you will likely come across several, because this sheep is gregarious and wants to be with others of its kind. The neat, alternating walking pattern is a direct or double register of the hind over the fore. When this sheep runs, its toes spread wide.

Similar Species: A Domestic Sheep's (*Ovis* spp.) print is similar. The introduced Barbarry Sheep (*Ammotragus lervia*), with a notable population in the Hondo Valley of New Mexico, also has similar prints. A Mule Deer (p. 18) print is more heart-shaped.

Feral Pig

**Fore and Hind Prints
(with dewclaws)**
Length: 2.5–3.0 in (6.4–7.6 cm)
Width: 2.3 in (5.8 cm)
Straddle
5.0–6.0 in (13–15 cm)
Stride
Trotting: 16–20 in (41–51 cm)
Size (male>female)
Height: 3.0 ft (91 cm)
Length: 4.3–6.0 ft (1.3–1.8 m)
Weight
77–440 lb (35–200 kg)

trotting

FERAL PIG (Wild Pig, Wild Boar)

Sus scrofa

Descended from introduced European animals, the Feral Pig interbred with escaped Domestic Pigs. Populations of this sturdy beast of the dense undergrowth can be found in small, scattered pockets throughout the region, notably in Arizona. Armed with tusks, they can be quite threatening.

A Feral Pig's print shows two prominent, widely spaced toe marks, and usually, except on firm surfaces, a clear, pointed dewclaw mark off to the side. The hind print is slightly smaller than the fore print. Feral Pigs are keen foragers, so their tracks can often be numerous, especially when they travel in a group. They usually travel in a trot, with a typical track showing a double register of hind over fore print, in the alternating pattern typically made by four-legged animals. Other signs are wallows and diggings.

Similar Species: Deer (pp. 16–21) tracks are similar, but with longer stride, pointier prints, a narrower gap between the toes and dewclaws to the rear, not the side. The Domestic Pig (also *Sus scrofa*) has a wider straddle and less neat tracks that often form two separate lines. Collared Peccary (p. 28) prints are much smaller and do not show dewclaws.

Collared Peccary

Fore Print
(hind print is slightly smaller)
Length: 0.8–1.5 in (2.0–3.8 cm)
Width: 0.8–1.5 in (2.0–3.8 cm)

Straddle
4.0–5.0 in (10–13 cm)

Stride
Walking: 6.5–10 in (17–25 cm)
Leaping: 6.0–10 ft (1.8–3.0 m)

Size
Height: 20–24 in (51–61 cm)
Length: 35–40 in (89–100 cm)

Weight
30–65 lb (14–29 kg)

walking

28

COLLARED PECCARY
(Javelina)
Tayassu tajacu

Stout and stocky, the coarse-furred peccary frequents the arid regions of southeastern Arizona and southern parts of New Mexico, where it feeds on the fleshy stems of cacti. Look near damaged cacti for tracks, and for diggings where peccaries have been sniffing out juicy roots. Piles of large, irregular scat and a musky or 'cheesy' smell are sure signs. Abroad only during the cooler parts of the day, it might bed down in the earth or in a cave during the hottest hours.

Each print shows two equal-sized toes; the dewclaws do not register. The hind prints are slightly smaller than the fore prints. An alternating pattern, with the hind print registering on or near the fore, is typical. Less aggressive than a Feral Pig, a peccary is capable of big leaps when startled.

Similar Species: The Feral Pig (p. 26), larger and less widespread, has prints with prominent dewclaws. Tracks of a young deer (pp. 16–21) may be mistaken for a peccary's.

Horse

**Fore Print
(hind print is slightly smaller)**
Length: 4.5–6.0 in (11–15 cm)
Width: 4.5–5.5 in (11–14 cm)

Stride
Walking: 17–27 in (43–69 cm)

Size
Height: to 6.0 ft (1.8 m)

Weight
to 1500 lb (680 kg)

walking

HORSE
Equus caballus

   This popular animal has unmistakable prints; it deserves mention because back-country use of the Horse means that you can expect its tracks to show up almost anywhere.

   The Horse has only one huge toe, which leaves an oval print. A distinctive feature is the 'frog' or V-shaped mark at the base of the print. When the Horse is shod, the horse-shoe shows up clearly as a firm wall at the outside of the print. Not all horses will be shod, so don't expect to see this outer wall on every horse track. A typical, leisurely horse track is an alternating walk with hind prints registering on or behind the slightly larger fore prints. Horses are capable of a range of speeds—up to a full gallop—but most recreational horseback riders prefer to walk their horses and soak up the mountain views!

Similar Species: Feral horses (also *Equus caballus*)—at large in New Mexico and Arizona and with the largest population in Nevada—have identical, unshod prints.

Black Bear

hind

fore

Fore Print
Length without heel:
　4.0–6.3 in (10–16 cm)
Width: 3.8–5.5 in (9.7–14 cm)

Hind Print
Length: 6.0–7.0 in (15–18 cm)
Width: 3.5–5.5 in (8.9–14 cm)

Straddle
9.0–15 in (23–38 cm)

Stride
Walking: 17–23 in (43–58 cm)

Size (male>female)
Height: 3.0–3.5 ft (91–110 cm)
Length: 5.0–6.0 ft (1.5–1.8 m)

Weight
200–600 lb (91–270 kg)

*walking
(slow)*

BLACK BEAR
Ursus americanus

Widespread in the Southwest's forested, mountainous parts, the Black Bear sleeps deeply in winter, so don't expect to encounter its tracks during the colder months. Finding fresh bear tracks can be a thrill, but take care, as the bear may be just around the corner. Never underestimate the potential power of a surprised bear!

A Black Bear's prints are about the length of a human's, but wider and with claw marks. The small inner toe rarely registers. The forefoot's small heel pad often shows, and the hind foot has a big heel. The bear's slow walk results in a slightly pigeon-toed double register of the hind print over the fore. At a faster pace, the hind oversteps the fore. When a bear runs, the two hind feet register in front of the forefeet in an extended cluster. Along well-worn bear paths, look for 'digs'–patches of dug-up earth–and 'bear trees' whose scratched bark shows that these bears climb.

Similar Species: Grizzly Bear (*Ursus arctos*) prints are similar, but this huge bear has been driven out of the region.

Coyote

fore

hind

Fore Print
(hind print is slightly smaller)
Length: 2.4–3.1 in (6.1–7.9 cm)
Width: 1.6–2.4 in (4.1–6.1 cm)

Straddle
4.0–7.0 in (10–18 cm)

Stride
Walking: 8.0–16 in (20–41 cm)
Galloping: 2.5–10 ft (76–300 cm)

Size (female is slightly smaller)
Height: 23–26 in (58–66 cm)
Length: 32–40 in (81–100 cm)

Weight
20–50 lb (9.1–23 kg)

walking | *gallop group*

COYOTE (Brush Wolf, Prairie Wolf)
Canis latrans

This widespread and adaptable canine prefers alpine meadows and open, grassy plains. It hunts rodents and larger prey, either on its own, with a mate or in a family pack. A Coyote also occasionally develops an interesting cooperative relationship with a Badger (p. 62), so you might find their tracks together where they have been digging for ground squirrels (p. 90).

The oval fore prints are slightly larger than the hind prints. Note the difference between the fore heel pad and the hind heel pad, which rarely registers clearly. Claw marks are usually evident only for the two center toes. A Coyote's tail hangs down, leaving a dragline in deep snow. Coyotes typically walk or trot in an alternating pattern—the walk has a wider straddle. When a Coyote gallops, the hind feet fall ahead of the forefeet; the faster it goes, the straighter the gallop group. A Coyote's trail is often direct, as if it knew exactly where it was going.

Similar Species: Domestic Dog (*Canis familiaris*) prints are not so oval and spread more, and a dog's trail is erratic and confused. Red Fox (*Vulpes vulpes*) prints are usually smaller.

Kit Fox

fore

hind

Fore and Hind Prints
Length: 1.1–1.8 in (2.8–4.6 cm)
Width: 1.1–1.5 in (2.8–3.8 cm)

Straddle
2.0–4.0 in (5.1–10 cm)

Stride
Walking/Trotting:
7.0–10 in (18–25 cm)

Size
Height: 12 in (30 cm)
Length with tail:
24–31 in (61–79 cm)

Weight
3.0–6.0 lb (1.4–2.7 kg)

trotting

KIT FOX (Swift Fox)

Vulpes velox

This small, shy fox hides away, on the plains and in desert regions of much of this region, avoiding the high mountains of central Arizona, Colorado and Utah. Secretive and solitary, its nocturnal activity makes it a rare sight, but evidence of its tracks can give it away. Its preference for sandy areas means that its tracks will seldom be very clear, as sand will often fall back into the print. If you find unclear prints, pay attention to more general track characteristics, such as whether the prints are in a typical dog-family trotting pattern. The track's dimensions may help distinguish which species made them.

Similar Species: A Gray Fox's (p. 38) track is very similar, but slightly larger. The Red Fox (*Vulpes vulpes*) has different heel pads (with a bar across them); in general, its prints are larger and less clear (because of thick fur), its stride longer and its straddle narrower. The prints of Domestic Cats (p. 46) and Bobcats (p. 44) lack claw marks and have a larger, less symmetrical heel pad.

Gray Fox

fore

hind

Fore Print
(hind print is slightly smaller)
Length: 1.3–2.1 in (3.3–5.3 cm)
Width: 1.1–1.5 in (2.8–3.8 cm)

Straddle
2.0–4.0 in (5.1–10 cm)

Stride
Walking/Trotting: 7.0–12 in (18–30 cm)

Size
Height: 14 in (36 cm)
Length: 21–29 in (53–74 cm)

Weight
7.0–15 lb (3.2–6.8 kg)

walking

GRAY FOX
Urocyon cinereoargenteus

Widespread in the Southwest, this small, shy fox especially prefers woodlands and brushy areas. It is the only fox that climbs trees, which it does either for safety or to forage.

The larger fore print registers better than the hind print, on which the long, semi-retractable claws do not always show. The heel pads are often unclear—they sometimes show up just as small, round dots. This fox has a neat, alternating walking track. When it trots, its prints fall in pairs, with the fore print set diagonally behind the rear print. Its gallop group is like a Coyote's (p. 34).

Similar Species: The smaller Kit Fox (p. 36), with similar (usually smaller) prints, prefers desert plains and other non-mountainous places. The Red Fox (*Vulpes vulpes*), with a more limited range, has heel pads with a bar across them; in general, its prints are larger and less clear (because of thick fur), its stride longer and its straddle narrower. The Domestic Cat (p. 46) and the Bobcat (p. 44) make similar trails, but their prints lacks claw marks and show a larger, less symmetrical heel pad.

Mountain Lion

fore

hind

Fore Print
(hind print is slightly smaller)
Length: 3.0–4.3 in (7.6–11 cm)
Width: 3.3–4.8 in (8.4–12 cm)

Straddle
8.0–12 in (20–30 cm)

Stride
Walking: 13–32 in (33–81 cm)
Bounding: to 12 ft (3.7 m)

Size
Height: 26–31 in (66–79 cm)
Length: 3.5–5.0 ft (1.1–1.5 m)

Weight
70–200 lb (32–91 kg)

walking (fast)

MOUNTAIN LION
(Puma, Cougar)
Felis concolor

The Mountain Lion is shy, elusive and nocturnal in nature, so finding its tracks is usually the best that trackers can hope for. Spread widely but sparsely because of its need for a big home territory, this large cat is essential in keeping the deer population down. It avoids large open areas, preferring mountainous terrain.

Mountain Lion prints tend to be wider than long and the retractable claws never register. In winter, thick fur makes the prints look much larger, and may obscure the two lobes on the front of the heel pad. When a Mountain Lion walks, the hind print will either directly register or double register on the larger fore print. As this cat's speed increases, the hind print tends to fall ahead of the fore print. In snow, the long, thick tail may leave a dragline, which can obscure some of the print detail. When necessary, the Mountain Lion is capable of bounding quickly to catch prey.

Similar Species: The beautiful, rare Jaguar (*Panthera onca*) of southeastern Arizona and southwestern New Mexico has nearly identical tracks. A Bobcat's (p. 44) prints are usually smaller, clearer and sunk into the snow more.

Jaguarundi

left fore

left hind

Fore Print
(hind print is slightly smaller)
Length: 1.5–1.8 in (3.8–4.6 cm)
Width: slightly less than length

Straddle
4.0–7.0 in (10–18 cm)

Stride
Walking: 7.0–10 in (18–25 cm)

Size
Length: 20–30 in (51–76 cm)

Weight
15–18 lb (6.8–8.2 kg)

running

JAGUARUNDI
Felis yagouaroundi

   This long-tailed, slender cat, which can be black, gray or reddish, has no distinctive markings. Confined to the extreme southeastern corner of Arizona, it is listed as an endangered species and it is a rare sight indeed. It favors forests and brushy areas with cacti and, although it spends most of its time on the ground, it can take to the trees to hunt and it easily swims rivers and ponds. Your best bet for tracks is in soft mud along the edges of waterbodies.

   The Jaguarundi's prints and tracks are typical for a cat. Its fore prints are slightly larger than its hind ones. Each foot registers four toes, but the retractable claws do not register. Though its short legs give it a short stride relative to its overall length, the Jaguarundi is extremely swift when pursuing prey.

Similar Species: The Mountain Lion (p. 40) and the Jaguar (*Panthera onca*) have larger prints. A juvenile Bobcat (p. 44) or a large Domestic Cat (p. 46) may have prints the same size. Though a fox (pp. 36–39) may leave similar tracks, its prints usually show claw marks.

Bobcat

fore

hind

**Fore Print
(hind print is slightly smaller)**
Length: 1.8–2.5 in (4.6–6.4 cm)
Width: 1.8–2.6 in (4.6–6.6 cm)

Straddle
4.0–7.0 in (10–18 cm)

Stride
Walking: 8.0–16 in (20–41 cm)
Running: 4.0–8.0 ft (1.2–2.4 m)

Size (female is slightly smaller)
Height: 20–22 in (51–56 cm)
Length: 25–30 in (64–76 cm)

Weight
15–35 lb (6.8–16 kg)

walking

*trotting
to loping*

BOBCAT (Wildcat)
Lynx rufus

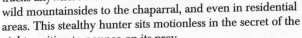

The Bobcat is
widely distributed
throughout the
Southwest. As Bobcats
are very adaptable
animals, you will
likely find their
tracks anywhere from
wild mountainsides to the chaparral, and even in residential
areas. This stealthy hunter sits motionless in the secret of the
night, waiting to pounce on its prey.

The hind print usually registers exactly on the larger
fore print in the walking track. Fore prints especially show
asymmetry. The heel pads have two lobes to the front and
three to the rear. As a Bobcat picks up speed, its track becomes a trot pattern made of groups of two prints, the hind
leading the fore. At even greater speeds, its track becomes a
group of four prints in a lope pattern. A Bobcat's feet leave
draglines in deep snow. Unlike the trails of wild dogs, the
Bobcat's trail meanders. Half-buried scat along the trail is a
sign of a Bobcat marking its territory.

Similar Species: The Ocelot (*Felis pardalis*), in brushy or
forested parts of southeastern Arizona, has similar prints
with less-scalloped heel pads. Jaguarundis (p. 42) and
juvenile Mountain Lions (p. 40) also leave similar prints.
A Domestic Cat (p. 46) has smaller prints, a shorter stride
and a narrower straddle. Coyote, dog and fox prints show
claw marks and have their footpads once-lobed at the front.

Domestic Cat

fore

hind

Fore Print
(hind print is slightly smaller)
Length: 1.0–1.6 in (2.5–4.1 cm)
Width: 1.0–1.8 in (2.5–4.6 cm)

Straddle
2.4–4.4 in (6.1–11 cm)

Stride
Walking: 5.0–8.0 in (13–20 cm)
Running: 14–32 in (36–81 cm)

Size (female is slightly smaller)
Height: 20–22 in (51–56 cm)
Length with tail: 30 in (76 cm)

Weight
6.5–13 lb (3.0–5.9 kg)

walking

*loping to
galloping*

DOMESTIC CAT
(House Cat)
Felis catus

The familiar Domestic Cat is so abundant in residential areas that its tracks can show up almost any place where there are people. From time to time, cats are abandoned and roam further afield. They are known as 'feral cats,' as they lead a pretty wild and independent existence. Domestic Cats can come in many shapes, sizes and colors.

As with all members of the cat family, a Domestic Cat's fore and hind prints both show four toe pads. Its retractable claws do not register–they are kept clean and sharp for catching prey. Cat prints usually show a slight asymmetry, with the toes pointed to one side. The hind print is slightly smaller than the fore print. A Domestic Cat makes a neat alternating walking track, usually in direct register, as one would expect from this animal's fastidious nature. When a cat picks up speed, it leaves clusters of four prints, the hind prints registering ahead of the fore prints.

Similar Species: A small Bobcat (p. 44) may leave tracks similar to a very large Domestic Cat's. Fox (pp. 36–39) and Domestic Dog (*Canis familiaris*) prints show claw marks.

Ringtail

Fore and Hind Prints
Length: 1.0–1.4 in (2.5–3.6 cm)
Width: 1.0–1.4 in (2.5–3.6 cm)
Straddle
3.0–4.0 in (7.6–10 cm)
Stride
Walking: 3.0–6.0 in (7.6–15 cm)
Size (female is slightly smaller)
Length: 24–32 in (61–81 cm)
Weight
1.5–2.5 lb (0.7–1.1 kg)

walking

RINGTAIL
(Cacomistle, Civet Cat, Miner's Cat)

Bassariscus astutus

This pretty cousin of the Raccoon is seldom seen, though it may be found in the foothills of most of the region's mountains and down into rocky canyons. The secretive Ringtail is strictly nocturnal and it rarely leaves any sign of its passage on the rocky terrain that it frequents. It usually travels under the cover of shrubs, adding to the difficulty of tracking this mammal. It never goes far from water.

A Ringtail's small, rounded prints show five toes; the partially retractable claws only occasionally register. Just behind the main pad of the fore print, a second pad may be evident. The common walking pattern is an alternating sequence of prints, where the hind registers on or close to the fore print. If you find a Ringtail's trail, it may lead you into rocky terrain, up a tree or to the animal's den.

Similar Species: Small weasel-family members (pp. 56–61, 64–67) have similar prints, but a Ringtail has a different gait and habitat, and its fifth toe registers more often. Domestic Cat (p. 46) prints never show five toes or a second pad.

Raccoon

fore

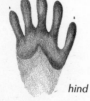

hind

Fore Print
Length: 2.0–3.0 in (5.1–7.6 cm)
Width: 1.8–2.5 in (4.6–6.4 cm)

Hind Print
Length: 2.4–3.8 in (6.1–9.7 cm)
Width: 2.0–2.5 in (5.1–6.4 cm)

Straddle
3.3–6.0 in (8.4–15 cm)

Stride
Walking/Running:
   7.0–20 in (18–51 cm)

Size (female is slightly smaller)
Length: 24–37 in (61–94 cm)

Weight
11–35 lb (5.0–16 kg)

walking   *running group*

RACCOON
Procyon lotor

The inquisitive Raccoon is adored for its distinctive face mask, yet disliked for its boundless curiosity–often demonstrated with garbage cans. Widely distributed in New Mexico and Colorado, it has a more scattered distribution in Arizona, Utah and Nevada–it avoids the driest parts. Look for its tracks near water in a diversity of habitats at lower elevations. Note that Raccoons like to rest in trees and that they usually den up for the colder months.

The Raccoon's unusual print looks like a human handprint, showing five well-formed toes. The small claws appear as dots. Its highly dexterous forefeet rarely leave heel prints, but its hind prints, which are generally much clearer, do show heels. The Raccoon's peculiar walking track shows a left fore print next to (or just ahead of) a right hind print and vice versa. On the rare occasions when a Raccoon is out in deep snow, it makes an alternating track. Raccoons occasionally run, leaving clusters in which the two hind prints fall ahead of the fore prints.

Similar Species: Unclear Opossum (p. 68) prints may look similar, but the Opossum drags its tail. The Fisher (*Martes pennanti*) prefers deeper forests and has different gaits.

White-nosed Coati

Fore and Hind Prints
Length: to 3.0 in (7.6 cm)
Width: to 2.0 in (5.1 cm)

Straddle
to 4.5 in (11 cm)

Stride
to 8.0 in (20 cm)

Size
(male>female)
Length with tail:
    33–53 in (84–130 cm)

Weight
17–27 lb (7.7–12 kg)

walking tracks of two coatis

WHITE-NOSED COATI
Nasua narica

 This inquisitive, sometimes-troublesome resident of southeastern Arizona and southwestern New Mexico is abroad by day, tail held high, in forested mountains or treed canyons. The attractive, furry coati responds to friendly gestures, especially if food is involved. However, bands of up to 25 animals will raid orchards for fruit, and when cornered, it will occasionally attack chickens or dogs.

 Both fore and hind prints show five toes; claws sometimes register on the fore print. The full heel registers only when a coati takes a rare rest. Males are solitary, but females and their young travel in groups, making individual tracks hard to distinguish. The White-nosed Coati is thought to be a newcomer from the tropical south and not many records of its tracks have been made yet.

Similar Species: The Mink (p. 58) and its large relatives may leave similar tracks, but their ranges do not overlap. The closely related Raccoon's (p. 50) prints are different.

River Otter

fore

hind

Fore Print
Length: 2.5–3.5 in (6.4–8.9 cm)
Width: 2.0–3.0 in (5.1–7.6 cm)

Hind Print
Length: 3.0–4.0 in (7.6–10 cm)
Width: 2.3–3.3 in (5.8–8.4 cm)

Straddle
4.0–9.0 in (10–23 cm)

Stride
Walking/Running: 6.0–23 in (15–58 cm)

Size
(female is slightly smaller)
Length with tail: 3.0–4.3 ft (91–130 cm)

Weight
10–25 lb (4.5–11 kg)

running (fast)

RIVER OTTER
Lutra canadensis

No animal knows how to have more fun than a River Otter. If you are lucky enough to see otters at play, you will not soon forget the experience. Well-adapted for the aquatic environment, this otter can be seen along streams and rivers throughout the Southwest, becoming increasingly rare towards the Mexican border. Within an otter's home territory, you are sure to find a wealth of evidence along the riverbanks.

In soft mud, the River Otter's five-toed feet show evidence of webbing, the hind prints more than the fore prints. The inner toes are set slightly apart. The heel may register, lengthening the print. Otter tracks are very variable: they show the typical two-print bounding of the weasel family, and, with faster runs, groups of four and three prints. The thick, heavy tail often leaves a mark over the prints. Otters love to slide in snow, often down riverbanks, leaving troughs nearly 12 inches (30 cm) wide. In summer they roll and slide on grass and mud.

Similar Species: Otter signs are quite distinctive, making identification easy.

Marten

Fore and Hind Prints
Length: 1.8–2.5 in (4.6–6.4 cm)
Width: 1.5–2.8 in (3.8–7.1 cm)

Straddle
2.5–4.0 in (6.4–10 cm)

Stride
Walking: 4.0–9.0 in (10–23 cm)
Running: 9.0–46 in (23–120 cm)

Size (male>female)
Length with tail:
    21–27 in (53–69 cm)

Weight
1.5–2.8 lb (0.7–1.3 kg)

walking   *bounding*

MARTEN (Pine Marten)
Martes americana

This aggressive predator is found in the coniferous montane and sub-alpine forests of northern New Mexico, Colorado and central Utah. Size and habitat are often key to identifying the Marten's tracks.

The Marten seldom leaves a clear print: the heel pad is very undeveloped and, in winter, the hairiness of the feet often obscures all pad detail, especially from the poorly developed palm pads. The heel pad is also very undeveloped. Often the inner toe fails to register, so the prints may show only four toes. In the Marten's alternating walking pattern, the hind print registers on the fore print. In a bounding track, the hind prints fall on the fore prints to form slightly angled print pairs, in the typical weasel-family pattern. Gallop groups may be three- or four-print clusters (see the River Otter, p. 54). If you follow the criss-crossing tracks, you may find that the Marten has scrambled up a tree; look for the sitzmark where it has jumped down again.

Similar Species: Large male Mink (p. 58) prints may be the same size as small female Marten prints, but Mink do not climb trees and, unlike Martens, are often found near water.

Mink

left fore

left hind

Fore and Hind Prints
Length: 1.3–2.0 in (3.3–5.1 cm)
Width: 1.3–1.8 in (3.3–4.6 cm)

Straddle
2.1–3.5 in (5.3–8.9 cm)

Stride
Walking/Running: 8.0–35 in (20–89 cm)

Size
(female is slightly smaller)
Length with tail: 19–28 in (48–71 cm)

Weight
1.5–3.5 lb (0.7–1.6 kg)

bounding

MINK
Mustela vison

The lustrous Mink, found in northern New Mexico, Colorado and scattered parts of Utah, prefers watery habitats surrounded by brush or forest. At home as much on land as in water, this nocturnal hunter can be an exciting animal to track. Like the River Otter (p. 54), the Mink likes to slide in snow, carving out a trough with its body for an observant tracker to spot.

The fore print of the Mink shows five (sometimes four) toes with five loosely connected palm pads in an arc, but the hind print shows only four palm pads. The heel pad rarely shows on the fore print but may register on the hind print, lengthening it. The Mink prefers the typical weasel-family bounding pattern of double prints, consistently spaced and slightly angled. Nevertheless, its tracks show much diversity in gait and may also appear as an alternating walk, or as a run with three- and four-print groups, as is illustrated for the River Otter.

Similar Species: Small Martens (p. 56), with similar prints, do not have a consistent double-print bounding gait or live near water. The Long-tailed Weasel (p. 60) has similar tracks.

Long-tailed Weasel

Fore and Hind Prints
Length: 1.1–1.8 in (2.8–4.6 cm)
Width: 0.8–1.0 in (2.0–2.5 cm)

Straddle
1.8–2.8 in (4.6–7.1 cm)

Stride
Bounding: 9.5–43 in (24–110 cm)

Size
(male>female)
Length with tail:
    12–22 in (30–56 cm)

Weight
3.0–12 oz (85–340 g)

bounding

LONG-TAILED WEASEL
Mustela frenata

Weasels are energetic hunters, with an avid appetite for rodents. Following their tracks can reveal much about these nimble creatures' activities. Although weasels are active all year, their tracks are most evident in winter. Look for holes where a weasel has suddenly plunged into deep snow to burrow beneath it, or perhaps to hunt rodents in their burrows. From time to time, weasel tracks may lead you up a tree. Weasels have also been known to take to water. Look for this weasel in southeastern Arizona, through most of New Mexico, Colorado and Utah.

Because of a weasel's light weight and small, hairy feet, the pad detail is often obscured, especially in snow. Even on clear prints, the inner (fifth) toe rarely registers. This weasel's tracks show typical bounding habits with an irregularity in the length of its stride—sometimes short and sometimes long—with no consistent behavior.

Similar Species: A large male Short-tailed Weasel (*Mustela erminea*), with a distribution from northern New Mexico into Colorado and central Utah, may leave tracks the same size as a small female Long-tailed Weasel. Mink (p. 58) tracks are similar.

Badger

fore

hind

walking

Fore Print
(hind print is slightly shorter)
Length: 2.5–3.0 in (6.4–7.6 cm)
Width: 2.3–2.8 in (5.8–7.1 cm)

Straddle
4.0–7.0 in (10–18 cm)

Stride
Walking: 6.0–12 in (15–30 cm)

Size
Length: 21–35 in (53–89 cm)

Weight
13–25 lb (5.9–11 kg)

BADGER
Taxidea taxus

Throughout the Southwest, the squat shape and unmistakable face of this bold animal are most likely to be seen in open grasslands, though the Badger also ventures into higher mountain country. Thick shoulders and forelegs, coupled with long claws, make for a powerful digging animal. In more northerly regions, unlike most other weasel-family members, the Badger likes to den up in a hole during the really cold months of winter. Look for its tracks there in spring and fall snow, though its wide, low body will often plow through deeper snow, obscuring print detail.

All five toes on each foot register. A Badger's long claws are evident in the pigeon-toed track that it leaves as it waddles along, although the claws on the hind feet are not as long as those on the forefeet. A Badger's alternating walking track is a double register, with the hind print sometimes falling just behind the fore print or just ahead of it.

Similar Species: In snow, a Porcupine's (p. 78) tracks may be similar, but will show draglines made by its tail and quills and it will likely lead up a tree, not to a hole.

Striped Skunk

fore

hind

Fore Print
Length: 1.5–2.2 in (3.8–5.6 cm)
Width: 1.0–1.5 in (2.5–3.8 cm)

Hind Print
Length: 1.5–2.5 in (3.8–6.4 cm)
Width: 1.0–1.5 in (2.6–3.8 cm)

Straddle
2.8–4.5 in (7.1–11 cm)

Stride
Walking/Running:
2.5–8.0 in (6.4–20 cm)

Size
Length with tail:
20–32 in (51–81 cm)

Weight
6.0–14 lb (2.7–6.4 kg)

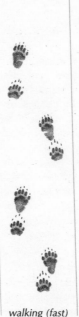

walking (fast)

running

STRIPED SKUNK
Mephitis mephitis

This striking skunk has a notorious reputation for its vile smell; the lingering odor is often the best sign of its presence. The widespread Striped Skunk enjoys a diversity of habitats throughout the Southwest.

Both fore and hind feet have five toes. The long claws on the forefeet often register. Smooth palm pads and small heel pads leave surprisingly small prints. Skunks mostly walk—with such a potent smell for their defence, and those memorable black and white stripes, they rarely need to run. Unlike other weasel-family members, skunks rarely show any consistent pattern in their tracks, but an alternating walking pattern may be evident. As a skunk speeds up, its hind print oversteps the fore. If a skunk runs, it leaves clumsy four-print groups. In snow it drags its feet.

Similar Species: The Hooded Skunk (*M. macroura*) of southeastern Arizona and southwestern New Mexico leaves identical prints. The Common Hognose (Rooter) Skunk (*Conepatus mesoleucus*), found in southern Arizona and from southern and New Mexico into eastern Colorado, has slightly longer toes. The Western Spotted Skunk (p. 66) leaves smaller prints in a very random pattern.

Western Spotted Skunk

fore

hind

Fore Print
Length: 1.0–1.3 in (2.5–3.3 cm)
Width: 0.9–1.1 in (2.3–2.8 cm)

Hind Print
Length: 1.2–1.5 in (3.0–3.8 cm)
Width: 0.9–1.1 in (2.3–2.8 cm)

Straddle
2.0–3.0 in (5.1–7.6 cm)

Stride
Walking: 1.5–3.0 in (3.8–7.6 cm)
Jumping: 6.0–12 in (15–30 cm)

Size
Length: 13–25 in (33–64 cm)

Weight
0.6–2.2 lb (0.3–1.0 kg)

walking

running

WESTERN SPOTTED SKUNK
(Civet Cat)
Spilogale gracilis

This beautifully marked skunk, smaller than its striped cousin, is distributed throughout the Southwest. It enjoys diverse habitats, such as scrubland, forests and farmland. It is a rare sight because of its nocturnal habits and because it dens up in winter, coming out only on warmer nights.

This skunk leaves a very haphazard trail as it forages for food on the ground. Occasionally, with ease, it climbs trees. Long claws on the forefeet often register, and the palm and heel may leave defined pad marks. Although this skunk rarely runs, when it does so it may bound along, leaving groups of four prints, hind ahead of fore. It sprays only when truly provoked, so its powerful odor is less frequently detected than that of the Striped Skunk (p. 64).

Similar Species: The larger Striped Skunk, with a similar range, has larger prints and less scattered tracks with a shorter running stride (or it jumps); it does not climb trees. A Hooded Skunk (*M. macroura*) leaves signs similar to a Striped Skunk's. The Hognose (Rooter) Skunk (*Conepatus mesoleucus*) has slightly longer toes.

Opossum

fore

hind

Fore Print
Length: 2.0–2.3 in (5.1–5.8 cm)
Width: 2.0–2.3 in (5.1–5.8 cm)

Hind Print
Length: 2.5–3.0 in (6.4–7.6 cm)
Width: 2.0–3.0 in (5.1–7.6 cm)

Straddle
4.0–5.0 in (10–13 cm)

Stride
5.0–11 in (13–28 cm)

Size
Length: 2.0–2.5 ft (61–76 cm)

Weight
9.0–13 lb (4.1–5.9 kg)

walking   *running*

OPOSSUM
Didelphis virginiana

This slow-moving nocturnal marsupial lives in the southeastern corner of Arizona, the valley of the Rio Grande in New Mexico, in northeastern Mexico and in pockets of Colorado. While the Opossum is found in many types of habitat, it shows a preference for open woodland or brushland around waterbodies. It is quite tolerant of farming and residential areas. Opossum tracks can often be seen in mud near the water; follow them and you might come across this strange animal playing dead ('playing 'possum') in the hope that you will leave it alone. If you find some roadkill, which the Opossum likes to feed on, look for tracks along the roadside.

Opossums are excellent climbers, so do not be surprised if their tracks lead to a tree. They have two walking habits: the common alternating pattern, with the hind prints registering on the fore prints, and a Raccoon-like paired-print pattern, with the hind print next to the opposing fore print. The very distinctive long inward-pointing thumb of the hind foot does not make a claw mark. Frostbite on the tail can sometimes result in little trails of blood in the snow.

Similar Species: Prints in which the distinctive thumbs don't show may be mistaken for a Raccoon's (p. 50).

Black-tailed Jackrabbit

fore

hind

Fore Print
Length: 1.5–3.0 in (3.8–7.6 cm)
Width: 1.3–1.7 in (3.3–4.3 cm)

Hind Print
Length: 2.5–4.0 in (6.4–10 cm)
Length with heel: to 6.0 in (15 cm)
Width: 1.5–2.5 in (3.8–6.4 cm)

Straddle
4.0–7.0 in (10–18 cm)

Stride
Hopping: 5.0–10 ft (1.5–3.0 m)
In alarm: 20 ft (6.1 m)

Size
Length: 18–25 in (46–64 cm)

Weight
4.0–8.0 lb (1.8–3.6 kg)

hopping

BLACK-TAILED JACKRABBIT
Lepus californicus

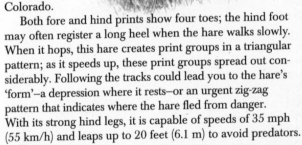

This seldom-seen nocturnal hare, often in groups, frequents open and agricultural areas throughout the Southwest, including desert country and into higher elevations, but not in the highest ranges of central New Mexico and Colorado.

Both fore and hind prints show four toes; the hind foot may often register a long heel when the hare walks slowly. When it hops, this hare creates print groups in a triangular pattern; as it speeds up, these print groups spread out considerably. Following the tracks could lead you to the hare's 'form'—a depression where it rests—or an urgent zig-zag pattern that indicates where the hare fled from danger. With its strong hind legs, it is capable of speeds of 35 mph (55 km/h) and leaps up to 20 feet (6.1 m) to avoid predators.

Similar Species: The Antelope Jackrabbit (*Lepus alleni*) lives in southern Arizona and New Mexico. The slightly larger White-tailed Jackrabbit (*Lepus townsendii*) lives in central Colorado and Utah. The Snowshoe Hare (p. 72), of more mountainous regions, spreads its hind toes more and takes shorter leaps. Coyote (p. 34) prints resemble heelless jackrabbit prints, but the gait is very different.

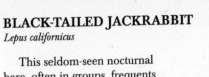

Snowshoe Hare

fore

hind

Fore Print
Length: 2.0–3.0 in (5.1–7.6 cm)
Width: 1.5–2.0 in (3.8–5.1 cm)
Hind Print
Length: 4.0–6.0 in (10–15 cm)
Width: 2.0–3.5 in (5.1–8.9 cm)
Straddle
6.0–8.0 in (15–20 cm)
Stride
Hopping: 0.8–4.2 ft (24–130 cm)
Size
Length: 12–21 in (30–53 cm)
Weight
2.0–4.0 lb (0.9–1.8 kg)

hopping

SNOWSHOE HARE (Varying Hare)
Lepus americanus

This hare is well known for its color change, from summer brown to winter white, and for its huge hind feet, which enable it to 'float' on the surface of snow. Widespread in the coniferous forests of the mountains of central New Mexico, Colorado and Utah, it favors brushy areas along creeks, both for food and to hide from the Coyote (p. 34), its most likely predator. Its well-worn runways serve as escape runs and it is most active at night.

As with rabbits and other hares, the Snowshoe Hare usually leaves a hopping track, with groups of four prints in a triangular pattern; they can be quite long if the hare is running quickly. The most distinctive feature of a hare's track is that the hind prints are much larger than the fore prints. In winter, heavy fur thickens the toes of a Snowshoe Hare's hind feet, which splay out to distribute its weight when it runs on snow. If you are lucky, you might even come across a resting hare, since the Snowshoe Hare does not live in burrows. A sign of this hare's presence is twigs and stems that have been neatly severed at a 45° angle.

Similar Species: A cottontail (p. 74) is smaller. Though jackrabbits (p. 70), in the open country, are larger, the Snowshoe Hare has larger hind prints.

73

Mountain Cottontail

hind

fore

Fore Print
Length: 1.0–1.5 in (2.5–3.8 cm)
Width: 0.8–1.3 in (2.0–3.3 cm)

Hind Print
Length: 3.0–3.5 in (7.6–8.9 cm)
Width: 1.0–1.5 in (2.5–3.8 cm)

Straddle
4.0–5.0 in (10–13 cm)

Stride
Hopping: 7.0–36 in (18–91 cm)

Size
Length: 12–17 in (30–43 cm)

Weight
1.3–3.0 lb (0.6–1.4 kg)

hopping

MOUNTAIN COTTONTAIL
(Nuttall's Cottontail)
Sylvilagus nuttallii

This abundant
rabbit is widespread
in the mountains
of northern New
Mexico and Arizona
and northward.
It prefers sagebrush,
but can also be found
in grassland and in rockier
areas, where it will hide in dense
vegetation and crevices to avoid predators.

As with other rabbits and hares, this rabbit's most common track is a triangular grouping of four prints. The larger hind prints (which can appear pointed) register ahead of the fore prints (which might merge together). The hairiness of the toes obscures any pad detail. If you follow this rabbit's tracks, you could be startled if it flies out from its 'form,' a depression in the snow or ground where it rests.

Similar Species: The similarly sized Desert (Audubon's) Cottontail (*S. audubonii*) and Eastern Cottontail (*S. floridana*) of southern Arizona and northern New Mexico prefer brushy desert country. Jackrabbits (p. 70) share the same range as the Mountain Cottontail, but leave much larger print clusters and take longer strides. Red Squirrel (p. 86) tracks have the fore prints more consistently side by side.

75

Pika

fore

hind

Fore Print
Length: 0.8 in (2.0 cm)
Width: 0.6 in (1.5 cm)

Hind Print
Length with heel:
  1.0–1.2 in (2.5–3.0 cm)
Width: 0.6–0.8 in (1.5–2.0 cm)

Straddle
2.5–3.5 in (6.4–8.9 cm)

Stride
Walking/Running: 4.0–10 in (10–25 cm)

Size
Length: 6.5–8.5 in (17–22 cm)

Weight
4.0–6.0 oz (110–170 g)

bounding

PIKA (Cony, Rock Rabbit)
Ochotona princeps

High up in the mountains of New Mexico, Colorado or Utah, you are more likely to hear the squeak of the Pika than to see it, as it is quick to disappear under the rocks for protection when alarmed. This cousin of the rabbits is confined to areas of high elevation. It rarely leaves good tracks because it prefers exposed, rocky areas on mountain slopes–its tracks are most likely to be found in spring, on patches of snow or in mud. A more conspicuous sign of the Pika's presence is its little hay piles, set to dry in the sun for the winter ahead, during which the Pika feeds on its stored food and remains active under the snow.

The Pika's fore print usually shows five toes, though one may not register. The hind print shows only four toes, and possibly the heel. Pika tracks may appear as an erratic alternating pattern or as three- and four-print bounding groups.

Similar Species: A rabbit or hare (pp. 70–75) may leave similar tracks, but has different habitats and never shows five toes on the fore prints. Moreover, the Pika, with smaller, rounder prints, does not have a long heel.

Porcupine

fore

hind

Fore Print
Length: 2.3–3.3 in (5.8–8.4 cm)
Width: 1.3–1.9 in (3.3–4.8 cm)

Hind Print
Length: 2.8–3.9 in (7.1–9.9 cm)
Width: 1.5–2.0 in (3.8–5.1 cm)

Straddle
5.5–9.0 in (14–23 cm)

Stride
Walking: 5.0–10 in (13–25 cm)

Size
Length with tail:
    2.2–3.4 ft (67–100 cm)

Weight
10–28 lb (4.5–13 kg)

walking

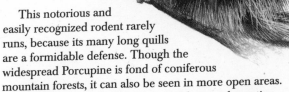

PORCUPINE
Erethizon dorsatum

This notorious and easily recognized rodent rarely runs, because its many long quills are a formidable defense. Though the widespread Porcupine is fond of coniferous mountain forests, it can also be seen in more open areas.

The most common Porcupine track is an alternating walking pattern, with the longer hind print registering on or slightly ahead of the fore print. Look for long claw marks on both prints. The fore print shows four toes and the hind print five. On clear prints, the unusual pebbly surface of the solid heel pads may show. However, a Porcupine's pigeon-toed footprints are often obscured by scratches from its heavy, spiny tail. A Porcupine's waddling gait shows in its track. In deeper snow, this squat animal drags its feet, and it may leave a trough with its body. A Porcupine's trail might lead you to a tree, where these animals spend much of their time feeding—look for chewed bark or nipped buds lying on the forest floor.

Similar Species: The Badger (p. 62) also has pigeon-toed prints, but it does not drag its tail or climb trees.

Beaver

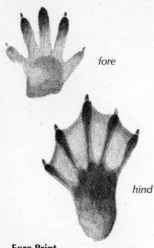

fore

hind

Fore Print
Length: 2.5–4.0 in (6.4–10 cm)
Width: 2.0–3.5 in (5.1–8.9 cm)

Hind Print
Length: 5.0–7.0 in (13–18 cm)
Width: 3.3–5.3 in (8.4–13 cm)

Straddle
6.0–11 in (15–28 cm)

Stride
Walking: 3.0–6.5 in (7.6–17 cm)

Size
Length with tail: 3.0–3.9 ft (91–120 cm)

Weight
28–75 lb (13–34 kg)

walking

BEAVER
Castor canadensis

Few animals leave as many signs of their presence as the Beaver, the largest North American rodent and a common sight around water. Its distribution largely follows watercourses; it is absent from arid places. Signs of Beaver activity include dams and domed lodges made of sticks and branches, and the stumps of felled trees—check trunks gnawed clean of bark bear marks of the Beaver's huge incisors. Beavers also make scent mounds marked with castoreum, a strong-smelling yellowish fluid.

A Beaver's tracks are often made less clear by its thick, scaly tail, or by the branches that it drags about for construction and food. If you find clear tracks, check the large hind prints for signs of webbing and broad toenails—the nail of the second inner toe usually does not show. Rarely do all five toes on each foot register. The Beaver's track may be in an irregular alternating sequence, often with a double register. Beavers frequently use the same path repeatedly, resulting in a well-worn trail.

Similar Species: Little confusion arises, because the Beaver leaves so many distinctive signs of its presence.

Muskrat

fore

hind

Fore Print
Length: 1.1–1.5 in (2.8–3.8 cm)
Width: 1.1–1.5 in (2.8–3.8 cm)

Hind Print
Length: 1.6–3.2 in (4.1–8.1 cm)
Width: 1.5–2.1 in (3.8–5.3 cm)

Straddle
3.0–5.0 in (7.6–13 cm)

Stride
Walking: 3.0–5.0 in (7.6–13 cm)
Running: to 1.0 ft (30 cm)

Size
Length with tail: 16–25 in (41–64 cm)

Weight
2.0–4.0 lb (0.9–1.8 kg)

walking

MUSKRAT
Ondatra zibethica

This rodent is found throughout the Southwest, but only where there is water, so it is absent from the arid southern parts of Arizona and New Mexico. Beavers (p. 80) are very tolerant of Muskrats and even allow them to live in parts of their lodges. Muskrats are active all year, and they leave plenty of signs of their presence. They dig an extensive network of burrows, often undermining the riverbank, so do not be surprised if you suddenly fall into a hidden hole! Other signs of this rodent are their small lodges in the water, and the beds of vegetation on which they rest, sun and feed during the summer.

The small inner toe of the five toes on each forefoot rarely registers in the print, but the hind print shows five well-formed toes. The prints are usually in an alternating track, with the hind print just behind or slightly overlapping the fore print. In snow, a Muskrat's feet drag a lot, and its tail leaves a sweeping dragline.

Similar Species: Few animals share this water-loving rodent's habits.

Yellow-bellied Marmot

fore

hind

Fore and Hind Prints
Length: 1.5–2.5 in (3.8–6.4 cm)
Width: 1.0–1.5 in (2.5–3.8 cm)

Straddle
3.0–5.0 in (7.6–13 cm)

Stride
Walking: 2.0–6.0 in (5.1–15 cm)
Running: 3.0–14 in (15–36 cm)

Size (male>female)
Length with tail:
17–28 in (43–71 cm)

Weight
5.0–10 lb (2.3–4.5 kg)

walking   *running*

YELLOW-BELLIED MARMOT

Marmota flaviventris

These endearing squirrels seem to have a good life—they sleep all winter and sunbathe on the rocks in summer. Yellow-bellied Marmots can be found high in the Rocky Mountains of New Mexico and Colorado, as well as in the mountain ranges of Utah. They live in small colonies, with an extensive network of burrows. Marmots are a joy to watch as they play-fight.

The fore print shows four toes and three palm pads; two heel pads may also be evident. The hind print shows five toes, four palm pads and two poorly registering heel pads. In a marmot's usual alternating walking pattern, its hind print registers over its fore print. When a marmot runs, its track shows a group of four prints, its two hind prints ahead of its fore prints. Because of this marmot's preference for rocky habitats, tracks may be hard to come by, but they can be found in spring and fall snowfalls.

Similar Species: A small Raccoon's (p. 50) running track may be confused with a Yellow-bellied Marmot's, but a Raccoon's fore print shows five toes; habitat and behavior are also good indicators.

Red Squirrel

fore

hind

Fore Print
Length: 0.8–1.5 in (2.0–3.8 cm)
Width: 0.5–1.0 in (1.3–2.5 cm)

Hind Print
Length: 1.5–2.3 in (3.8–5.8 cm)
Width: 0.8–1.3 in (2.0–3.3 cm)

Straddle
3.0–4.5 in (7.6–11 cm)

Stride
Running: 8.0–30 in (20–76 cm)

Size
Length with tail:
9.0–15 in (23–38 cm)

Weight
2.0–9.0 oz (57–260 g)

bounding

*bounding
(in deep snow)*

86

RED SQUIRREL
(Pine Squirrel, Chickaree)
Tamiasciurus hudsonicus

When you enter its
territory, a Red Squirrel
greets you with a
loud, chattering call.
Another obvious sign of
this squirrel, whose range
includes the mountains of
central New Mexico and
Arizona and extends northwards into Utah and Colorado,
is the large middens—piles of cone
scales and cores—at the bases of trees
that indicate favorite feeding sites.

Active all year in their small territories, Red Squirrels
make an abundance of tracks that lead from tree to tree
and down burrows. These energetic animals mostly run,
with a gait that leaves groups of four prints; the five-toed
hind prints fall ahead of the four-toed fore prints, which
are often side by side, and the heels often do not register.
In deeper snow, a Red Squirrel's prints merge to form
diamond-shaped pairs.

Similar Species: Tracks in a similar pattern, but with
smaller straddles and smaller prints, were probably made
by a chipmunk (p. 92) or, in coniferous forests at higher
altitudes in southern and central Utah, the Northern Flying
Squirrel (*Glaucomys sabrinus*). A cottontail's (p. 74) fore prints
rarely register side by side when it runs.

Black-tailed Prairie Dog

fore

hind

Hind Print
(fore print is slightly smaller)
Length: to 1.3 in (3.3 cm)
Width: to 0.8 in (2.0 cm)

Straddle
to 3.0 in (7.6 cm)

Size
Length with tail: 14–17 in (36–43 cm)

Weight
2.0–3.0 lb (0.9–1.4 kg)

running

BLACK-TAILED PRAIRIE DOG

Cynomys ludovicianus

From southeastern Arizona through New Mexico and the plains of eastern Colorado, the Black-tailed Prairie Dog was once a familiar sight, until poisoning reduced its numbers. Gradually making a comeback across its former territory, this highly social animal can be a joy to watch in its 'prairie dog towns,' which may house several thousand residents. It is hard to miss when it barks out warnings to its neighbors that a naturalist is approaching.

Fore prints show four toes and hind ones show five, one of which resembles a small thumb. Palm pads are quite evident, especially on the hind print. Look for tracks in the mud around its burrows and excavations. These prairie dogs are so busy that it is hard to find a track that has not been overrun. Notice the raised rim of earth around the burrow entrance, which serves the resident as a lookout.

Similar Species: Gunnison's Prairie Dog (*C. gunnisoni*) lives in more diverse habitats in the Four Corners area.
The Utah Prairie Dog (*C. parvidens*) of southwestern Utah leaves comparable tracks, as does the Rock Squirrel (p. 90).

Rock Squirrel

fore

hind

Fore Print
Length: 1.0–1.5 in (2.5–3.8 cm)
Width: 0.7–1.2 in (1.8–3.0 cm)

Hind Print
Length: 2.1–2.5 in (5.3–6.4 cm)
Width: 1.0–1.7 in (2.5–4.3 cm)

Straddle
2.0–5.0 in (5.0–13 cm)

Stride
Running: 7.0–18 in (18–46 cm)

Size
Length with tail: 17–21 in (43–53 cm)

Weight
21–28 oz (600–790 g)

running

ROCK SQUIRREL
Spermophilus variegatus

The Rock Squirrel, the largest, most widespread of the many ground squirrels in the Southwest, enjoys mountain slopes, even into deep canyons. In the milder south it remains active all year. Look for tracks in mud around the many burrow entrances and, in northern parts, in late or early snowfalls, before or after hibernation.

The small fifth toe and the two heel pads of the forefoot rarely register. The larger hind print shows five toes. The long claws of both fore and hind feet frequently register. Ground squirrels leave typical squirrel tracks, with the hind prints registering ahead of the diagonally placed fore prints.

Similar Species: The Golden Mantled Ground Squirrel (*S. lateralis*) occupies mountainous regions. The petite Spotted Ground Squirrel (*S. spilosoma*) inhabits New Mexico, eastern Arizona and eastern Colorado. The Thirteen-lined Ground Squirrel (*S. tridecemlineatus*) lives in northern New Mexico and in Colorado. The much smaller White-tailed Antelope Squirrel (*Ammospermophilus leucurus*) of northern Arizona, Utah and Nevada has similar tracks. A Red Squirrel's (p. 86) tracks have a more square running group.

Colorado Chipmunk

fore

hind

Fore Print
Length: 0.8–1.0 in (2.0–2.5 cm)
Width: 0.4–0.8 in (1.0–2.0 cm)

Hind Print
Length: 0.7–1.4 in (1.8–3.6 cm)
Width: 0.5–0.9 in (1.3–2.3 cm)

Straddle
2.0–3.1 in (5.1–7.9 cm)

Stride
Running: 7.0–15 in (18–38 cm)

Size
Length with tail: 8.1–9.4 in (21–24 cm)

Weight
2.0 oz (57 g)

bounding

COLORADO CHIPMUNK
Tamias quadrivittatus

Of the many delightful chipmunks in the Southwest, one of the more widespread is the Colorado Chipmunk, a dark-colored inhabitant of ponderosa pine and other forests in New Mexico. It has wider tastes in habitat in Colorado and also occurs in northeastern Arizona and southeastern Utah.

Chipmunks are so light that their prints rarely show fine details. They run on their toes, so their forefoot heel pads often do not register; the hind feet have no heel pads. The hind feet have five toes and register ahead of the four-toed forefeet. A chipmunk's erratic trail often leads to extensive burrows, or occasionally up a tree, where it went to forage.

Similar Species: The Gray-footed Chipmunk (*T. canipes*) lives in southeastern New Mexico. The Gray-collared Chipmunk (*T. cinereicollis*) inhabits east-central Arizona and southwestern New Mexico. The Cliff Chipmunk (*T. dorsalis*) occurs in central Arizona and into Utah and Nevada. The Least Chipmunk (*T. minimus*) lives in high mountains and forest. Palmer's Chipmunk (*T. palmeri*) lives in southern Nevada. The Red Squirrel (p. 86) leaves larger prints and remains active through mountain winters while chipmunks hibernate. Mice (p. 104–109) prints are smaller.

White-throated Woodrat

fore

hind

Fore Print
Length: 0.6–0.8 in (1.5–2.0 cm)
Width: 0.4–0.5 in (1.0–1.3 cm)
Hind Print
Length: 1.0–1.5 in (2.5–3.8 cm)
Width: 0.6–0.8 in (1.5–2.0 cm)
Straddle
2.3–2.7 in (5.8–6.9 cm)
Stride
Walking: 1.8–3.0 in (4.6–7.6 cm)
Jumping: 5.0–8.0 in (13–20 cm)
Size
Length with tail:
    11–16 in (28–41 cm)
Weight
4.8–10 oz (140–280 g)

walking   *running*

WHITE-THROATED WOODRAT
Neotoma albigula

This nocturnal woodrat thrives in arid regions through-out New Mexico, most of Arizona and into southern Colorado. The trail of a White-throated Woodrat might lead you to a distinctive mass of a nest, most often under a prickly-pear or cholla cactus. This woodrat relishes spiny meals too; it gets most of its water from cacti.

Four toes register on the fore print and five on the hind. The short claws rarely show. A woodrat often walks in an alternating fashion, with the hind print directly registering on the fore print. This woodrat also frequently runs, leaving a pattern of four prints, with the larger hind prints register-ing ahead of the diagonally placed fore prints. Its stride tends to be short relative to the size of its feet.

Similar Species: The Bushy-tailed Woodrat (*N. cinerea*) inhabits northern Arizona and New Mexico and into Utah and Colorado. The Mexican Woodrat (*N. mexicana*) lives in eastern Arizona and western New Mexico. The Southern Plains Woodrat (*N. micropus*) is found in New Mexico. The juniper-dependent Stephen's Woodrat (*N. stephensi*) oc-curs in central Arizona and New Mexico. The Norway Rat (p. 96) is usually found close to human activity. The Yellow-bellied Marmot (p. 84) has similar, much larger prints.

Norway Rat

fore

hind

Fore Print
Length: 0.7–0.8 in (1.8–2.0 cm)
Width: 0.5 in (1.3 cm)

Hind Print
Length: 1.0–1.3 in (2.5–3.3 cm)
Width: 0.8–1.0 in (2.0–2.5 cm)

Straddle
3.0 in (7.6 cm)

Stride
Walking: 1.5–3.5 in (3.8–8.9 cm)
Jumping: 5.0–12 in (13–30 cm)

Size
Length with tail:
    13–19 in (33–48 cm)

Weight
7.0–18 oz (200–510 g)

walking

NORWAY RAT
(Brown Rat)
Rattus norvegicus

This despised rat is widespread almost anywhere humans have decided to build homes, although it is not entirely dependent on people and may be found in the wild as well. It is active both day and night.

The fore print of the Norway Rat shows four toes and the hind print shows five; the hind heel does not show. The Norway Rat commonly leaves an alternating walking track pattern, with the larger hind print registering close to or on the fore. When it runs, this colonial rat leaves groups of four prints with the diagonally placed fore prints registering behind the hind prints. In snow, the rat's tail often leaves a dragline. Since rats live in groups, you may find that there are many tracks close together, often leading to their 2-inch (5.1 cm) wide burrows in the ground.

Similar Species: Woodrat (p. 94) tracks may be similar, but woodrats rarely associate with human activity, except in abandoned buildings. The Roof (Black) Rat (*Rattus rattus*), as widespread as the Norway Rat, is smaller.

Hispid Cotton Rat

fore

hind

Fore Print
Length: 0.5–0.7 in (1.3–1.8 cm)
Width: 0.5–0.7 in (1.3–1.8 cm)

Hind Print
Length: 0.6–1.0 in (1.5–2.5 cm)
Width: 0.6–0.8 in (1.5–2.0 cm)

Straddle
1.3–1.5 in (3.3–3.8 cm)

Stride
Walking: 1.3 in (3.3 cm)

Size
Length with tail: 8.0–14 in (20–36 cm)

Weight
2.8–7.0 oz (79–200 g)

walking

HISPID COTTON RAT

Sigmodon hispidus

The Hispid Cotton Rat, with a range from southeastern Arizona into New Mexico and Colorado, makes itself unpopular by eating valuable crops. Though keen on eating almost anything green, it prefers grassy fields. It stays close to home and consequently its little runways clearly mark its routes to favored feeding sites.

The fore prints show four toes while the larger hind prints usually show five. The heel of the hind foot will not always register, especially if the rat is moving fast. This medium-sized rodent leaves a typical walking track in which the hind print registers on and slightly behind the fore print. Watch for its nests, which are balls of woven grass, and small piles of cut grass.

Similar Species: The Arizona Cotton Rat (*S. arizonae*) is found only in southern Arizona. The Tawny-bellied Cotton Rat (*S. fulviventer*) lives in central-southern New Mexico and southeastern Arizona. The Yellow-nosed Cotton Rat (*S. ochrognathus*) inhabits extreme southeastern Arizona and southwestern New Mexico. The Norway Rat (p. 96) makes similar tracks but is usually found close to human activity. Woodrat (p. 94) tracks are similar but with wider straddle.

Botta's Pocket Gopher

fore

hind

Fore Print
Length: 1.0 in (2.5 cm)
Width: 0.6 in (1.5 cm)

Hind Print
Length: 1.0–1.5 in (2.5–3.8 cm)
Width: 0.5 in (1.3 cm)

Straddle
1.5–2.0 in (3.8–5.1 cm)

Stride
Walking: 1.3–2.0 in (3.3–5.1 cm)

Size (male>female)
Length with tail: 6.5–11 in (17–28 cm)

Weight
2.5–8.0 oz (71–230 g)

walking

BOTTA'S POCKET GOPHER

Thomomys bottae

This seldom-seen rodent can be found throughout the Southwest, except for the eastern margin of New Mexico. It spends most of its life in burrows, but from time to time it will venture out to move soil about or to find a mate. Pocket gophers prefer soft and sandy or moist soils, but can be found even in very dry habitats.

By far the best sign of Botta's Pocket Gopher is its muddy mounds and tunnel cores–they are especially evident just after spring thaw. Each mound marks the entrance to a burrow, which is always blocked up with a plug. Search around the mounds to find tracks. Both fore and hind feet have five toes. The forefeet have well-developed long claws for digging, though it is a rare track that shows this much detail. The typical track is an alternating walking pattern, where the hind print registers on or slightly behind the fore print.

Similar Species: The Northern Pocket Gopher (*T. talpoides*) lives in northern parts of Arizona and New Mexico and in Colorado and Utah. The Plains Pocket Gopher (*Geomys bursarius*) and the Mexican Pocket Gopher (*Cratogeomys castanops*) both inhabit eastern New Mexico.

Mexican Vole

fore

hind

Fore Print
Length: 0.5 in (1.3 cm)
Width: 0.5 in (1.3 cm)

Hind Print
Length: 0.6 in (1.5 cm)
Width: 0.5–0.8 in (1.3–2.0 cm)

Straddle
1.3–2.0 in (3.3–5.1 cm)

Stride
Walking: 0.8 in (2.0 cm)
Running/Hopping:
    2.0–6.0 in (5.1–15 cm)

Size
Length with tail:
    6.3–8.4 in (16–21 cm)

Weight
1.5–3.0 oz (43–85 g)

walking

hopping (in snow)

102

MEXICAN VOLE
Microtus mexicanus

There are so many vole species that positive track identification is next to impossible. The Mexican Vole, the region's most widespread vole, is commonly found anywhere from boggy areas to yellow-pine forest clearings to dry grassy areas. Its range extends from southeastern Utah down into Mexico, but it is absent from western Arizona and eastern New Mexico.

Vole fore prints show four toes and hind prints five, but they are seldom clear. In a vole's paired alternating walking track, the hind print may register on the fore; it always does so in a vole's preferred hopping gait. Voles stay under the snow in winter so, when the snow melts, look for distinctive piles of cut grass from their ground nests and for tiny teeth marks in the bark at shrub bases. In summer, vole paths become little runways through the grass.

Similar Species: The Southern Red-backed (*Clethrionomys gapperi*) and Montane (*M. montanus*) voles inhabit the Rocky Mountains. The Long-tailed Vole (*M. longicaudus*) lives in northern Arizona, much of New Mexico and northwards. Deer Mouse (p. 106) tracks show a paired hop pattern.

Silky Pocket Mouse

hind

fore

slow hopping group (in sand)

Fore Print
Length: 0.3 in (0.8 cm)
Width: 0.3 in (0.8 cm)
Hind Print
Length: 0.6 in (1.5 cm)
Width: 0.4 in (1.0 cm)
Straddle
1.5 in (3.8 cm)
Stride
Running: 0.8–4.5 in (2.0–11 cm)
Size
Length with tail: 3.9–4.8 in (9.9–12 cm)
Weight
0.2–0.3 oz (6–9 g)

bounding

SILKY POCKET MOUSE
Perognathus flavus

A large number of endearing pocket mice can be found scattered throughout the Southwest. The nocturnal Silky Pocket Mouse, the most widespread one in the region, is found in most of New Mexico, in eastern Arizona and in Colorado. It prefers to forage in thinly vegetated rocky or sandy areas. Though it may hibernate during winter cold spells, in milder areas it remains active all year.

As a pocket mouse bounds along, it leaves four-print groups: the two fore prints fall closely side by side and the larger, wider-set hind feet overstep them. The trail is seldom clear. Look for the little depressions in which pocket mice take dust baths to get rid of ticks and mites.

Similar Species: The Plains Pocket Mouse (*P. flavescens*) shares the same range. The Hispid Pocket Mouse (*Chaetodipus hispidus*) lives in southeastern Arizona, eastern New Mexico and Colorado. The Rock Pocket Mouse (*C. intermedius*) and the Desert Pocket Mouse (*C. penicillatus*) inhabit southern Arizona and New Mexico. Deer Mouse (p. 106) tracks are similar but slightly larger.

Deer Mouse

running group

Fore Print
Length: 0.3 in (0.8 cm)
Width: 0.3 in (0.8 cm)

Hind Print
Length: 0.6 in (1.5 cm)
Width: 0.4 in (1.0 cm)

Straddle
1.4–1.8 in (3.6–4.6 cm)

Stride
Running: 2.0–5.0 in
(5.1–13 cm)

Size
Length with tail:
6.0–9.0 in (15–23 cm)

Weight
0.5–1.3 oz (14–37 g)

running

*running
(in snow)*

DEER MOUSE
Peromyscus maniculatus

One of many mice in the Southwest, this highly adaptable rodent might be encountered almost anywhere, except at the highest elevations. It may enter buildings and remain active in winter.

When clear, Deer Mouse fore prints each show four toes, three palm pads and two heel pads, and the hind prints show five toes and three palm pads; the hind heel pads rarely register. Running tracks—most noticeable in snow—show the hind prints in front of the close-set fore prints. The tracks may lead you up a tree or down into a burrow. In soft snow, the prints may merge and appear as larger pairs of prints, with tail drag evident.

Similar Species: Grasshopper mice (*Onychomys* spp.) and harvest mice (*Reithrodontomys* spp.) have similar tracks. The Brush Mouse (*P. boylii*) prefers the brushland of lower mountains. The Piñon Mouse (*P. truei*) lives among piñon and juniper. The Canyon Mouse (*P. crinitus*) and Cactus Mouse (*P. eremicus*) also have similar tracks. The House Mouse (*Mus musculus*) is more associated with humans. A vole's (p. 102) merged two-print pattern has longer strides. Jumping mice (p. 108) may have similar tracks. Chipmunks (p. 92) have a wider straddle. Shrews (p. 112) have a narrower straddle.

Western Jumping Mouse

jumping group

Fore Print
Length: 0.3–0.5 in (0.8–1.3 cm)
Width: 0.3–0.5 in (0.8–1.3 cm)

Hind Print
Length: 0.5–1.3 in (1.3–3.3 cm)
Width: 0.5 in (1.3 cm)

Straddle
1.8–1.9 in (4.6–4.8 cm)

Stride
Hopping: 2.0–7.0 in (5.1–18 cm)
In alarm: 3.0–4.0 ft (91–120 cm)

Size
Length with tail: 7.0–9.0 in (18–23 cm)

Weight
0.6–1.3 oz (17–37 g)

jumping

WESTERN JUMPING MOUSE
Zapus princeps

Congratulations if you find and successfully identify the tracks of a Western Jumping Mouse! This hard-to-find rodent can be found in east-central Arizona, through parts of New Mexico and into Colorado and Utah. The Western Jumping Mouse's preference for lush tall-grass meadows and its long, deep hibernation period (up to six months!) make locating tracks very difficult.

Jumping mouse tracks are distinctive if you do find them. The two smaller fore prints register between the long hind feet. The long heels do not always show. When they jump, these mice make short leaps. The tail may leave a dragline in soft mud or unseasonable snow. An abundant sign of this rodent is the clusters of cut grass stems, about 5 inches (13 cm) long, lying in the meadows.

Similar Species: A Deer Mouse (p. 106) track may have the same straddle. A kangaroo rat (p. 110) also jumps, but on its hind feet only, not on all four.

Ord's Kangaroo Rat

slow hop group

Hind Print
(fore print is much smaller)
Length: 1.5–1.8 in (3.8–4.6 cm)
Width: 0.5–0.8 in (1.3–2.0 cm)
Straddle
1.3–2.3 in (3.3–5.8 cm)
Stride
Hopping: 5.0–24 in (13–61 cm)
Size
Length with tail: 8.0–14 in (20–36 cm)
Weight
1.5–2.5 oz (43–71 g)

fast hop

ORD'S KANGAROO RAT
Dipodomys ordi

This small, athletic rodent is capable of big jumps, as its name suggests. Look for this rat in the shrubby arid regions of New Mexico, in all but the southwest of Arizona and into Utah and at lower elevations in Colorado. In cold weather it stays under the snow, venturing out on milder nights.

Though good tracks are hard to find, you may see prints in sand, but without fine detail; habit best identifies the track. When a kangaroo rat hops slowly, its two small forefeet register between its large hind feet, which show long heel marks, and its long tail leaves a dragline. At speed, the fore prints do not register, the hind heels appear shorter and the tail registers infrequently. Kangaroo rats can have either four or five toes per hind foot. If you find the large nesting mounds of this kangaroo rat, tap your fingers beside a burrow and you may hear thumping in reply.

Similar Species: The Desert Kangaroo Rat (D. *deserti*) lives in southwestern Arizona and southern Nevada. Merriam's Kangaroo Rat (D. *merriami*) extends further into southern New Mexico. The Banner-tailed Kangaroo Rat (D. *spectabilis*) inhabits southern Arizona and much of New Mexico. The Western Jumping Mouse (p. 108) prefers lush areas and leaves smaller tracks.

Desert Shrew

running group

Fore Print
Length: 0.2 in (0.5 cm)
Width: 0.2 in (0.5 cm)

Hind Print
Length: to 0.4 in (1.0 cm)
Width: to 0.3 in (0.8 cm)

Straddle
0.8–1.3 in (2.0–3.3 cm)

Stride
Running: 1.2–2.0 in (3.0–5.1 cm)

Size
Length with tail: 3.0–3.6 in (7.6–9.1cm)

Weight
to 0.2 oz (6 g)

running

DESERT SHREW
Notiosorex crawfordi

Of the many tiny, frenetic shrews in the Southwest, the Desert Shrew, widespread throughout Arizona and New Mexico and northward, is one of the most likely. It is especially fond of sagebrush and prickly-pear cactus habitat, but its rapid activity makes it difficult to observe closely.

In its energetic and unending quest for food, a shrew usually leaves a running pattern of four prints that may slow to an alternating walking pattern. The individual prints in a group are often indistinct, but in mud or shallow, wet snow, you can even count the five toes on each print. In snow, a shrew's tail often leaves a dragline. If the shrew tunnels under the snow, it may leave a ridge of snow on the surface. A shrew's trail may disappear down a burrow.

Similar Species: The Wandering Shrew (*Sorex vagrans*) inhabits northeastern Arizona, western New Mexico, Utah and Colorado. The larger Water Shrew (*S. palustris*) is often found by cold mountain streams. The tiny Dwarf Shrew (*S. nanus*) and the Dusky Shrew (*S. monticolus*) live at higher altitudes. Merriam's Shrew (*S. merriami*) can be found from west-central Arizona to west-central New Mexico and northwards. Mice (pp. 104–109) fore prints show four toes.

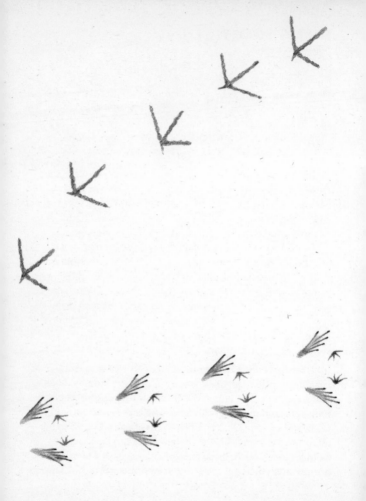

BIRDS, AMPHIBIANS AND REPTILES

A guide to the animal tracks of Arizona and New Mexico is not complete without some consideration of the birds, amphibians and reptiles in the region. Only a few have been chosen as examples.

The differences among bird species are not necessarily reflected in their tracks, so several bird species have been chosen to represent the main types common to this region.

Bird tracks can often be found in abundance in snow and are clearest in shallow, wet snow. The shores of streams and lakes are very reliable locations to find bird tracks—the mud there can hold a clear print for a long time. The sheer number of tracks made by shorebirds and water-fowl can be astonishing. While some bird species prefer to perch in trees or soar across the sky, it can be entertaining to follow the tracks of those that do spend a lot of time on the ground. They can spin around in circles and lead you in all directions. The track may suddenly end as the bird takes flight, or it might terminate in a pile of feathers, the bird having fallen victim to a hungry predator.

Many amphibians and turtles depend on moist environments, so look in the soft mud along the shores of lakes and ponds for their distinctive tracks. While you may be able to distinguish frog tracks from toad tracks, since they generally move differently, it can be very difficult to identify the species. In drier environments, reptiles outnumber amphibians, but dry terrain does not show prints well, except in sand. Snakes can leave distinctive tracks as they wind their way through mud or sand; foot-less, they make body prints.

Dark-eyed Junco

Print
Length: to 1.5 in (3.8 cm)
Straddle
1.0–1.5 in (2.5–3.8 cm)
Stride
Hopping: 1.5–5 in (3.8–13 cm)
Size
5.5–6.5 in (14–17 cm)

DARK-EYED JUNCO

Junco hyemalis

This common small bird typifies the many small hopping birds found in this region. Each foot has three forward-pointing toes and one longer toe at the rear. The best prints are left in snow, although in deep snow the toe detail is lost, and the feet may show some dragging between the hops.

A good place to study these types of prints is near a birdfeeder. Watch the birds scurry around as they pick up fallen seeds, then have a look at the prints that they have left behind.

Similar Species: The size of the toes may indicate what kind of bird you are tracking–larger birds have larger footprints. Not all birds are present year-round, so keep in mind the season when tracking. A similar alternating track in open fields is likely from the Horned Lark (*Eremophila alpestris*), a year-round resident of the Southwest that prefers to run rather than hop.

Great Horned Owl

Strike
Width to 3.0 ft (91 cm)
Size
22 in (56 cm)

GREAT HORNED OWL
Bubo virginianus

This wide-ranging owl is often seen resting quietly in trees during the day, as it prefers to hunt at night. It is an accomplished hunter in snow, and the 'strike' that it leaves can be quite a sight if it registers well. The owl strikes through the snow with its talons, leaving an untidy hole, which is occasionally surrounded by imprints of wing and tail feathers. These feather imprints are made as the owl struggles to take off with possibly heavy prey. It is not the most graceful of walkers, preferring to fly away from the scene.

You might stumble across this owl's strike and guess that its target could have been a vole scurrying around underneath the snow. If you are a really lucky tracker, you will have been following the surface track of an animal to find that it abruptly ends with this strike mark, where the animal has been seized by an owl.

Similar Species: The Common Raven (*Corvus corax*) also leaves strike marks, usually with much sharper feather imprints.

Mallard

Print
Length: 2.0 to 2.5 in (5.1–6.4 cm)
Straddle
4.0 in (10 cm)
Stride
to 4.0 in (10 cm)
Size
23 in (58 cm)

MALLARD
Anas platyrhynchos

female

male

This dabbling duck–the male a familiar sight with its striking green head–is common in open areas by lakes and ponds through much of the Southwest. Its prints can often be seen in abundance along the muddy shores of just about any waterbody, including those in urban parks.

The webbed foot of the Mallard has three long toes that all point forward. Though the toes register well, the webbing between them does not always show on the print. The inward-pointing feet give the Mallard a pigeon-toed appearance, which perhaps accounts for its waddling gait, a characteristic for which ducks are known.

Similar Species: The many gulls and the many dabblers, such as the warm-colored Cinnamon Teal (*Anas cyanoptera*) and the elegant Northern Pintail (*A. acuta*), are examples of the many waterfowl with similar prints. Exceptionally large webbed prints are likely from a goose, such as the Canada Goose (*Branta canadensis*) or, in southern New Mexico, perhaps a winter-visiting Tundra Swan (*Cygnus columbianus*).

Spotted Sandpiper

Print
Length: 0.8–1.3 in (2.0–3.3 cm)
Straddle
to 1.5 in (3.8 cm)
Stride
Erratic
Size
7.0–8.0 in (18–20 cm)

SPOTTED SANDPIPER

Actitis macularia

The bobbing tail of the Spotted Sandpiper is a common sight on the shores of lakes, rivers and streams, but you will usually find only one of these territorial birds in any given location. Because of its excellent camouflage, likely the first you will see of this bird will be when it flies away, its fluttering wings close to the surface of the water.

As it teeters up and down on the shore, it leave trails of three-toed prints. Its fourth toe is very small and faces off to one side at an angle. Sandpiper tracks can have an erratic stride.

Similar Species: All sandpipers and plovers, including the common Killdeer (*Charadrius vociferus*), have similar tracks, although there is much diversity in print size.

Great Blue Heron

Print
Length: to 6.5 in (17 cm)
Straddle
8.0 in (20 cm)
Stride
9.0 in (23 cm)
Size
4.2–4.5 ft (1.3–1.4 m)

GREAT BLUE HERON
Ardea herodias

The refined and graceful image of this large heron symbolizes the precious wetlands in which it patiently hunts for food. Usually still and statuesque as it waits for a meal to swim by, the Great Blue Heron will walk from time to time, leaving precise prints in a nearly straight line along the banks or mudflats of waterbodies. Note the slender rear toe.

Similar Species: Cranes (*Grus* spp.) have similar habitats and prints, but their smaller rear toes do not usually register. Bitterns (*Botaurus* spp.), ibises (*Plegadis* spp.) and rails (*Rallus* spp.) are waterfowl with similar tracks. The Common Moorhen (*Gallinula chloropus*) of southern Arizona and New Mexico has enormous feet (compared to its body) that allow it to stay on top of deep mud or floating vegetation.

Roadrunner

Print
Length: 3.0 in (7.6 cm)
Straddle
Standing: 4.0 in (10 cm)
Stride
(varies with speed) to 12 in (30 cm)
Size
23 in (58 cm)

ROADRUNNER
Geococcyx californianus

   The Roadrunner
is a resident of large,
open desert regions and
plains from southern Nevada
across to eastern New Mexico. Like the popular cartoon
character of the same name, the Roadrunner spends most
of its time speeding across the plains on its strong legs.
Using its keen eyesight and tough bill, it hunts and eats
insects, small reptiles, rodents and even other birds.

   A Roadrunner's foot has two forward-pointing toes and
two much larger rearward-pointing ones. The four-pointed
star shape of the print is unmistakable, even in sand. The
Roadrunner hardly ever flies; its preference for the ground
and its love of dry areas can result in a mass of tracks.

Similar Species: Few tracks are like those of this ground-
loving member of the cuckoo family. The Northern Flicker
(*Colaptes auratus*) has a similar arrangement of toes, but its
print is only about 1.8 inches (4.6 cm) long.

Frogs

Straddle
to 3.0 in (7.6 cm)

FROGS

Bullfrog

Look for frog
tracks along the
muddy fringes
of waterbodies.
Frogs generally hop, leaving
symmetrical groups of prints in which two large, long-toed
hind prints fall outside of two smaller fore prints.

The aridity of much of the region limits frog diversity.
Two widespread frogs whose tracks you might find are the
Rio Grande Leopard Frog (*Rana berlandieri*), which favors
wet spots in southwestern Arizona and extreme southern
New Mexico, and the Northern Leopard Frog (*R. pipiens*),
more to the north, in parts of New Mexico and into Utah
and Colorado, in wet areas from the lowlands to the high-
lands. Both can reach 5 inches (13 cm) in length.

The diminutive toad-like Canyon Treefrog (*Hyla areni-
color*), to just 2.3 inches (5.8 cm) in length, likes canyon
streams. It is found through most of Arizona and into western New Mexico and southern Utah, always close to water.
Slightly smaller still is the Burrowing Treefrog (*Pternohyla
fodiens*), found only in south-central Arizona. It adapts to
the dry conditions by burrowing into the soil.

An unusually large track is surely from the robust
Bullfrog (*R. catesbeiana*). At up to 8 inches (20 cm) in length,
this giant, introduced from the east, is spreading its range
through many parts of the Southwest. It will always be close
to ponds and streams, often resting on the bank.

Toads

Straddle
to 3.0 in (7.6 cm).

TOADS

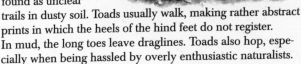

Woodhouse's
Toad

Toad tracks,
best sought along
the muddy fringes
of waterbodies,
might also be
found as unclear
trails in dusty soil. Toads usually walk, making rather abstract
prints in which the heels of the hind feet do not register.
In mud, the long toes leave draglines. Toads also hop, especially when being hassled by overly enthusiastic naturalists.

Toads are frequently better adapted than frogs to the
arid parts of the Southwest. For example, the spadefoots
burrow into loose soils of plains and river basins to escape
the dryness—look for a small mound of soil surrounded by
toad prints. The most widespread spadefoot is the Western
Spadefoot (*Scaphiopus hammondi*), which is found in eastern
Arizona and throughout New Mexico. It can grow up to
2.5 inches (6.4 cm) in length.

The widespread Red-spotted Toad (*Bufo punctatus*), to
3 inches (7.6 cm) long, can be found far from water, sometimes in rodent burrows. Woodhouse's Toad (*B. woodhousei*),
found near wet areas in New Mexico, Arizona and Colorado, can grow to 5 inches (13 cm) long. Almost as large,
the Great Plains Toad (*B. cognatus*) prefers the grasslands and
plains of New Mexico and southern Arizona.

The plain-colored Colorado River Toad (*B. alvarius*), up
to 7 inches (18 cm) in length, lives in moist pockets of the
desert regions of southwestern Arizona.

Lizards and Salamanders

Lizards

LIZARDS AND SALAMANDERS

Eastern Fence Lizard

The variety of dry terrain of the Southwest is perfectly suited for many different species of long, slender lizard-like reptiles, making track identification difficult. Also, tracks may be made less clear by a dragging belly or a thick tail.

The attractively marked Banded Gecko (*Coleonyx variegatus*), common in rocky or sandy parts of southern and western Arizona, hides by day to avoid the intense heat.

The Eastern Fence Lizard (*Sceloporus undulatus*), to over 9 inches (23 cm) long, lives in many different habitats, except the deserts of southwestern Arizona and Nevada. The stout and colorful Gila Monster (*Heloderma suspectum*) is a highly venomous lizard of southern and central Arizona and southern Nevada. It may move into burrows made by other animals or, especially in moist soils, it makes its own.

Of the sleek and swift skinks, the most likely candidate is the widespread Many-lined Skink (*Eumeces multivirgatus*), which is absent only from southern Nevada. It can grow to 7.5 inches (19 cm) in length and can often be found near rocks in the grassy plains and deserts.

Salamanders, which need moister environments, have similar shapes and similar tracks, but with shorter toe marks. The Tiger Salamander (*Ambystoma tigrinum*) lives in many habitats, from the plains to the mountains, in New Mexico, Colorado and Utah.

TURTLES AND TORTOISES

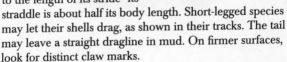

Desert Tortoise

With its large shell and short legs, a turtle leaves a track that is wide relative to the length of its stride—its straddle is about half its body length. Short-legged species may let their shells drag, as shown in their tracks. The tail may leave a straight dragline in mud. On firmer surfaces, look for distinct claw marks.

Since most turtles enjoy marshes and ponds, they are not plentiful in this region. Nevertheless, along the muddy edges of slow-moving streams and rivers in the plains of eastern New Mexico and in the extreme south of Arizona, you might find the tracks of the nocturnal Yellow Mud Turtle (*Kinosternon flavescens*), which grows to 6.5 inches (17 cm) long. Similar in size but more widespread, the Sonora Mud Turtle (*K. sonoriense*) lives in southern Arizona and in the extreme southwest of New Mexico.

The smaller Western Box Turtle (*Terrapene carolina*) clambers around the open prairies and woodlands of New Mexico and southwestern Arizona. Look closely at cattle dung for the tracks of this turtle, which eats dung beetles. Larger turtle-type tracks in arid parts of western Arizona and southern Nevada among creosote bushes and cacti were likely made by the slow, stocky Desert Tortoise (*Gopherus agassizii*), which can grow to a length of 14 inches (36 cm). During the heat of the day it hides in its burrow.

Snakes

typical snake track

Sidewinder track

SNAKES

Sidewinder

Many species of
snakes thrive in the
arid Southwest, espe-
cially in southwestern
Arizona and southern
Nevada. Soft sand and
riverside mud are among
the best places for snake tracks.

One of the more widespread snakes in the Southwest,
though absent from northern New Mexico and much of
Colorado, is the small and secretive Night Snake (*Hypsi-
glena torquata*), which grows to only 26 inches (66 cm) in
length. The Common Kingsnake (*Lampropeltis getulus*),
which enjoys a variety of habitats throughout the region,
can reach a length of 6.8 feet (2.1 m). Sturdy and strong,
and found in most of the Southwest, the Pine-gopher Snake
(*Pituophis melanoleucus*) can grow to 8.3 feet (2.5 m) long.

The most widespread rattlesnake, found anywhere from
coast to timberline, but not in southwestern Arizona, is
the adaptable Western Rattlesnake (*Crotalus viridis*), which
grows to 5.3 feet (1.6 m) long. For trackers, the most distinctive rattler, to 33 inches (84 cm) in length, is the swift, agile
Sidewinder (*Crotalus cerastes*), which leaves J-shaped marks.
Other snake tracks are so similar that identification of the
species is next to impossible (consider habitat and range)
and even the direction of travel can be hard to determine.

FOSSILIZED TRACKS

Scattered throughout the Southwest are many sites where there are fossilized tracks. Millions of years ago, strange animals roamed the land, leaving tracks just as animals do today. People have found the fossilized tracks not only of huge dinosaurs and tiny ones, but also of ancient birds, dogs and cats and even of primitive frogs.

However, it was the really massive dinosaurs who left the most impressive tracks. Standing in a print up to 33 inches (84 cm) long made seventy million or more years ago brings history to life as you imagine a massive *Tyrannosaurus rex* squelching its way along a muddy shore.

Track of a bipedal dinosaur (one that ran about on its hind legs only), such as Tyrannosaurus rex.

Track of a quadrupedal dinosaur (one that walked about on all fours), such as a brontosaur or Triceratops.

As you leave your own tracks along a lakeshore, consider that they too may become fossilized and that one day in the far future they could be discovered by some other advanced organism with a passion for tracking history!

TRACK PATTERNS & PRINTS

Elk
p. 16

Mule Deer
p. 18

White-tailed Deer
p. 20

Pronghorn
p. 22

Bighorn Sheep
p. 24

Feral Pig
p. 26

TRACK PATTERNS & PRINTS

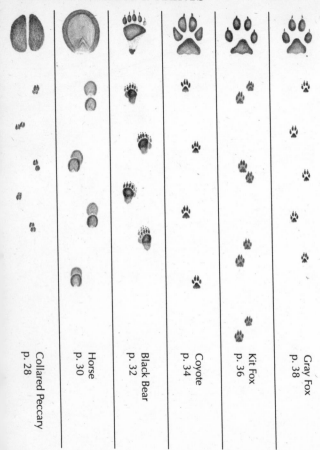

Collared Peccary
p. 28

Horse
p. 30

Black Bear
p. 32

Coyote
p. 34

Kit Fox
p. 36

Gray Fox
p. 38

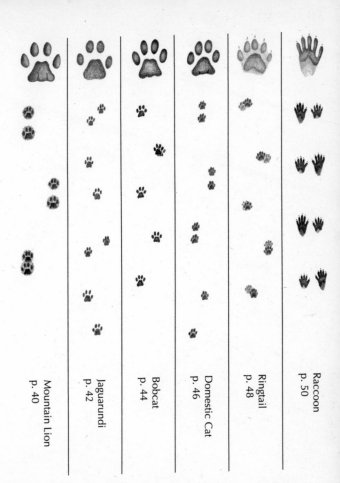

Mountain Lion
p. 40

Jaguarundi
p. 42

Bobcat
p. 44

Domestic Cat
p. 46

Ringtail
p. 48

Raccoon
p. 50

TRACK PATTERNS & PRINTS

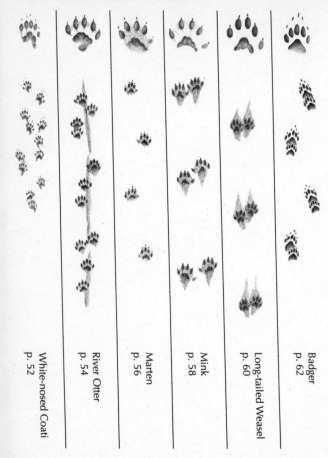

White-nosed Coati
p. 52

River Otter
p. 54

Marten
p. 56

Mink
p. 58

Long-tailed Weasel
p. 60

Badger
p. 62

142

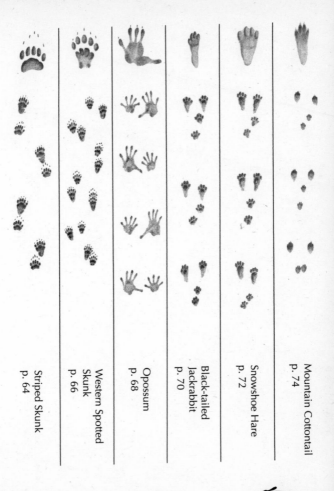

Striped Skunk
p. 64

Western Spotted
Skunk
p. 66

Opossum
p. 68

Black-tailed
Jackrabbit
p. 70

Snowshoe Hare
p. 72

Mountain Cottontail
p. 74

143

TRACK PATTERNS & PRINTS

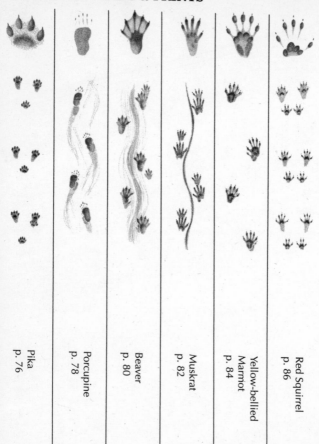

Pika
p. 76

Porcupine
p. 78

Beaver
p. 80

Muskrat
p. 82

Yellow-bellied
Marmot
p. 84

Red Squirrel
p. 86

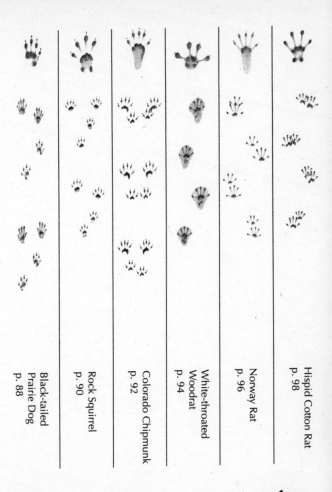

Black-tailed
Prairie Dog
p. 88

Rock Squirrel
p. 90

Colorado Chipmunk
p. 92

White-throated
Woodrat
p. 94

Norway Rat
p. 96

Hispid Cotton Rat
p. 98

145

TRACK PATTERNS & PRINTS

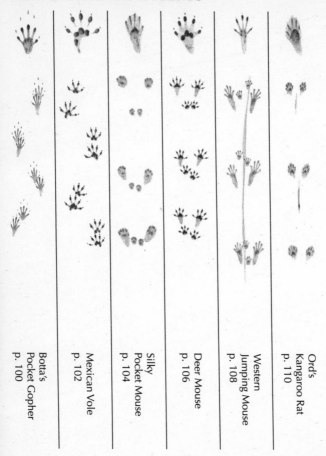

Botta's
Pocket Gopher
p. 100

Mexican Vole
p. 102

Silky
Pocket Mouse
p. 104

Deer Mouse
p. 106

Western
Jumping Mouse
p. 108

Ord's
Kangaroo Rat
p. 110

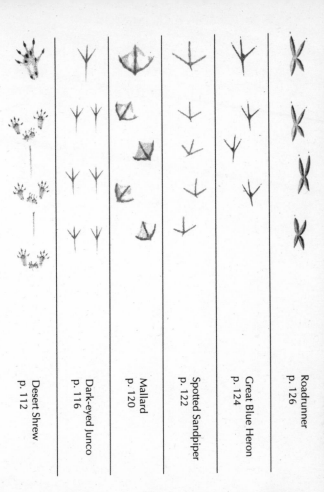

Desert Shrew
p. 112

Dark-eyed Junco
p. 116

Mallard
p. 120

Spotted Sandpiper
p. 122

Great Blue Heron
p. 124

Roadrunner
p. 126

TRACK PATTERNS & PRINTS

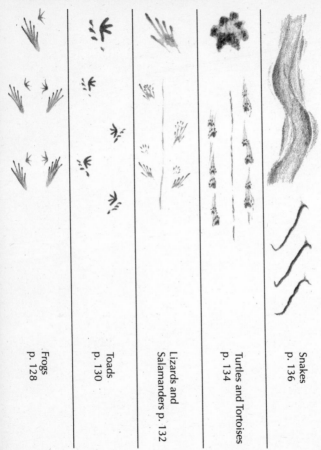

Frogs
p. 128

Toads
p. 130

Lizards and
Salamanders p. 132

Turtles and Tortoises
p. 134

Snakes
p. 136

HOOFED PRINTS

Collared Peccary

Mule Deer

Feral Pig

White-tailed Deer

Bighorn Sheep

Pronghorn
Antelope

Elk

Horse

149

FORE PRINTS

Ringtail

Western Spotted Skunk

Long-tailed Weasel

Mink

Striped Skunk

Marten

White-nosed Coati

Badger

River Otter

Domestic Cat

Kit Fox

Jaguarundi

Gray Fox

Bobcat

Coyote

Mountain Lion

Black Bear

inch | cm
0 | 0
1 |
2 | 5

HIND PRINTS

Norway Rat

Ord's Kangaroo Rat

Red Squirrel

Yellow-bellied Marmot

Rock Squirrel

Muskrat

Opossum

Raccoon

Porcupine

Mountain Cottontail

Snowshoe Hare

Black-tailed Jackrabbit

Beaver

```
inch    cm
0       0

        1

2       5
```

151

HIND PRINTS

**Desert
Shrew**

**Deer
Mouse**

**Mexican
Vole**

**Hispid
Cotton Rat**

Pika

**Silky
Pocket Mouse**

**Western
Jumping Mouse**

**Black-tailed
Prairie Dog**

**Colorado
Chipmunk**

**Botta's
Pocket Gopher**

**White-throated
Woodrat**

BIBLIOGRAPHY

Behler, J.L. and F.W. King. 1979. *Field Guide to North American Reptiles and Amphibians.* National Audubon Society. New York: Alfred A. Knopf, Inc.

Burt, W.H. 1976. *A Field Guide to the Mammals.* Boston: Houghton Mifflin Company.

Farrand, J., Jr. 1995. *Familiar Animal Tracks of North America.* National Audubon Society Pocket Guide. New York: Alfred A. Knopf, Inc.

Forrest, L.R. 1988. *Field Guide to Tracking Animals in Snow.* Harrisburg: Stackpole Books.

Halfpenny, J. 1986. *A Field Guide to Mammal Tracking in North America.* Boulder: Johnson Publishing Company.

Headstrom, R. 1971. *Identifying Animal Tracks.* Toronto: General Publishing Company, Ltd.

Lockley, M.G. 1995. *Dinosaur Tracks and Other Fossil Footprints of the Western United States.* New York: Columbia University Press.

Murie, O.J. 1974. *A Field Guide to Animal Tracks.* The Peterson Field Guide Series. Boston: Houghton Mifflin Company.

Rezendes, P. 1992. *Tracking and the Art of Seeing: How to Read Animal Tracks and Sign.* Vermont: Camden House Publishing, Inc.

Stall, C. 1990. *Animal Tracks of the Southwest.* Seattle: The Mountaineers.

Stokes, D. and L. Stokes. 1986. *A Guide to Animal Tracking and Behaviour.* Toronto: Little, Brown and Company.

Whitaker, J.O., Jr. 1996. *National Audubon Society Field Guide to North American Mammals.* New York: Alfred A. Knopf, Inc.

INDEX

Page numbers in **boldface** type refer to the primary
(illustrated) treatments of animal species and their tracks.

ABOUT THE AUTHOR

Ian Sheldon has lived in South Africa, Singapore, Britain and Canada. Caught collecting caterpillars at the age of three, he has been exposed to the beauty and diversity of nature ever since. He was educated at Cambridge University, England, and the University of Alberta. When he is not in the tropics working on conservation projects or immersing himself in our beautiful wilderness, he is sharing his love for nature. An accomplished artist, naturalist and educator, Ian enjoys communicating passion through the visual arts and the written word, in the hope that he will inspire love and affection for all nature.